D0064801

DESERTIFICATION IN EUROPE

Proceedings of the Information Symposium in the EEC Programme on Climatology,
held in Mytilene, Greece, 15–18 April 1984,
under the sponsorship of the Commission of the European Communities
Directorate-General for Science, Research and Development

UNESCO/MAB

University of Thessaloniki

Governmental and Local Hellenic Authorities

Publication arrangements

P. P. ROTONDÓ

Commission of the European Communities
Directorate-General Information Market and Innovation

Commission of the European Communities

Desertification
in
Europe

Proceedings of the Information Symposium
in the EEC Programme on Climatology,
held in Mytilene, Greece, 15-18 April 1984

Edited by

R. FANTECHI

Commission of the European Communities,
Directorate-General for Science, Research and Development, Brussels, Belgium

and

N. S. MARGARIS

University of Thessaloniki, Division of Ecology,
Department of Biology, School of Sciences, Thessaloniki, Greece

D. REIDEL PUBLISHING COMPANY

A MEMBER OF THE KLUWER ACADEMIC PUBLISHERS GROUP

DORDRECHT / BOSTON / LANCASTER / TOKYO

Library of Congress Cataloging in Publication Data

Information Symposium in the EEC Programme on Climatology (1984: Mytiléné,
   Lesbos Island, Greece)
   Desertification in Europe.

   At head of title: Commission of the European Communities.
   Includes index.
     1. Desertification–Europe–Congresses. 2. Desert soils–Europe–Con-
gresses. I. Fantechi, Roberto, II. Margaris, N. S. III. Commission of the
European Communities. IV. Title.
GB648.42.I54   1984   551.4   86–3290
ISBN 90–277–2230–7

Publication arrangements by
Commission of the European Communities
Directorate-General Information Market and Innovation, Luxembourg

EUR 10395
© 1986 ECSC, EEC, EAEC, Brussels and Luxembourg

LEGAL NOTICE
Neither the Commission of the European Communities nor any person acting on behalf of the
Commission is responsible for the use which might be made of the following information.

Published by D. Reidel Publishing Company
P.O. Box 17, 3300 AA Dordrecht, Holland

Sold and distributed in the U.S.A. and Canada
by Kluwer Academic Publishers,
190 Old Derby Street, Hingham, MA 02043, U.S.A.

In all other countries, sold and distributed
by Kluwer Academic Publishers Group,
P.O. Box 322, 3300 AH Dordrecht, Holland

Printed in The Netherlands

Despite the fact that Spain already finds itself on the UNO map of desertification, when mentioning the subject of desertification in Europe the most usual reaction is well that embodied in the title of Prof. Mensching's paper : "Desertification in Europe ?", and according to the various interlocutors the question mark may express any nuance from candid astonishment to perplexity and beyond. Most of the problem concerning the acceptance of the concept of "Desertification in Europe" is semantic in character. "Desert" as a place where sand dunes roam about with hardly any vegetation and no water is the mental picture one generally gets. Yet "desertification" is no desert : it is, as the suffix -ication clearly indicates, a _process_, of which a full desert is one end, a healthy ecosystem being the other. It is a degradation process which deserves being studied in any of its phases because of the dangers involved. When the cost of the restoration of a land to its former productivity is higher than whatever net gain may have been obtained at the price of the decrease in productivity, one has to acknowledge that some precious resources have been lost, irrespective of the fact that rain may fall and certain plants still grow. Something has to be abandoned, deserted, and that is desertification.

The definition of "desertification" has been the object of many discussions. Normally, and quite understandably, the various definitions found in the literature reflect more their authors' concern and scientific interest rather than an inquiry into the actual meaning of the word itself. We would venture to suggest that Rapp's definition ("the spread of desert-like conditions in arid and semi-arid areas up to 600 mm, due to man's influence or to climatic change") is the one which best fits the concept we are discussing. The rainfall limit set by Rapp is easily seen to include most of Southern Europe. The human component is the key to the observed difference in conditions between, say, Greece and Sahel. The climatic component is a warning that man's activities may in the end be unable to avoid passing the point beyond which the process is no longer reversible.

The papers collected in this book provide evidence that the continuous loss of fertile soil is a constant in at least the Mediterranean Europe. The problem is not new. As the Symposium on European Desertification took place in Greece, it is becoming to evoke that greatest among men, Plato[1]. In **Critias**, 111, b-d, he gives a picture of ancient Attica. He says that "many and remarkable floods have occurred ...... The mass of soil which descended steeply from high places did not expand, as it does elsewhere, in terraces, and its great flow rolling down unceasingly was finally lost in the deep. Since then what remains, as one can see it in small islands, offers, when comparing the present conditions to those then prevailing, the image of a body that a sickness has made skeletal, once all that

----
[1] I owe to A.T. Grove that my attention has been brought upon this passage of Plato.

which earth has of fat and soft has left the bones, only the
fleshless body of the land remaining". Plato goes on recalling
the time when the country "had vast forests on its mountains ...
which today only have what can nourish bees ... The land was
carrying abundant pastures for cattle. Moreover, ... the water
which every year fell from sky to fecundate them, was not lost by
the soil as it is today and did not leave the bare ground to flow
down into the sea; on the contrary the soil had plenty of it and
stored it in its bosom and kept it under a mantle of clay. Then
water, descending from high places, ... provided everywhere an
inexhaustible flow to fountains and rivers".

Here is a precise account of both climatic and anthropogenic
factors of land degradation, as we say nowadays. A climate
change can also be inferred from Plato's description, and one may
argue whether man has had a part in it.

Today, man's role in environmental degradation is increasing in
importance. As A.T. Grove points out, we have to face
accelerated erosion and sedimentation, soil salinization, falling
water-tables, further deterioration of the plant cover and urban
sprawl, processes which, even though they operate locally, have
nevertheless a Community-wide dimension. If one views all those
processes in the frame of a semi-arid climate, and if one is
willing to consider the possibility of a man-induced climate
change in a short or medium term, due to increasing atmospheric
CO_2, the need for a lucid awareness of dangers and opportunities,
such as sound, applied research can provide, is only too obvious.

It is therefore with satisfaction that the editors acknowledge
the support and encouragement received from Dr. Ph. Bourdeau,
Director, Environment, Raw Materials and Materials Technologies
Research Programmes (DG XII), Commission of the European
Communities, Brussels.

Roberto Fantechi

C O N T E N T S

A. DESERTIFICATION : FACTORS AND PROCESSES

Desertification in Europe? A critical comment with examples from Mediterranean Europe

The scale factor in relation to the processes involved in "Desertification" in Europe

Desertification in a changing climate with a particular attention to the Mediterranean countries

The effects of model-generated climatic changes due to a CO_2 doubling on desertification processes in the Mediterranean area

Landscape changes in Greece as a result of changing climate during the quaternary

Comparison of climatic evolution during post-glacial times in Greece, tropical and subtropical regions, in relation to desertification

Climatic implications of glacier fluctuations

DESERTIFICATION IN EUROPE ?

A Critical Comment With Examples From Mediterranean Europe

H.G. MENSCHING
Chairman of the IGU Working Group
Resource Management in Drylands
University of Hamburg, F.R.G.

Summary

A seminar on "Desertification in the European Countries"
must point out clearly what the term "Desertification"
means and where it applies. Desertification is not simply
the process of land degradation or deterioration of natural ecological conditions in any region or country.
Desertification is specifically defined by the way human
impact can produce desert-like conditions in the affected
areas. This is only possible where regions are located in
climatic zones which are characterized by a definite
degree of climatic aridity. In Europe this is only the
case in countries with a Mediterranean climate and sufficient summer dryness (cp. for example "Carte Bioclimatique
de la Région Méditerranéenne", UNESCO-FAO, 1962). It is
only in these areas that 'desert-like conditions' can
occur locally or regionally. In the humid countries, the
degradational consequences following from human impact
differ distinctly from those produced by desertification
processes. Examples of consequences following from
desertification are given from Spain and Sicily.

1. Introduction

     Aridity combined with high variability of rainfall from
year to year and season to season is the mother of desertification. Any excessive form of human impact on a semi-arid ecosystem, however, is its father!
     Such semi-arid regions are wide-spread in Mediterranean
southern Europe. On the Iberian Peninsula they reach far north
up to the southern slopes of the Pyrenees. In Italy they include the whole Mezzogiorno with the islands of Sardinia and
Sicily. Almost all of Greece lies within this ecologically
endangered zone, enhanced by the circumstance that a great part
of its area is taken in by limestone massifs.
     Semi-arid ecosystems require all of their rainy season,
i.e. winter rains in this case, to render possible the restoration of the water balance indispensable for the regeneration of
the vegetation cover and soil fertility and stability (against
processes of erosion for example). The longer and more excessive this period of regeneration is interrupted or impeded by
man, the severer the results from human impact will be: first
through a general degradation of the natural potential, then -
if no protective measures are introduced - phenomena of desert-

ification and finally irreversible soil destruction will follow.
    For more than 2,000 years the Mediterranean countries have
belonged to the most intensely used and exploited parts of
Europe. Especially during the antique Greek and Roman imperi-
alist periods exploitation of the natural ecological potential
reached a serious degree in various regions and caused severe
desertification. One of the reasons was the ruthless deforestation practiced in wide parts of the Iberian, Roman and Greek
Mediterranean. The landscape deterioration that resulted is
still obvious today.

2. The Historical Process of Desertification

    Mediterranean Europe is characterized by a very old histo-
ry of cultivation. Intensive land use began many centuries
before Christ. As the torrential winter rains often flooded the
valleys and coastal plains and these were infected by malaria,
cultivated areas concentrated on the hill and mountain slopes.
The cultivation of cereals, especially wheat, required extensive clearing and deforestation, coinciding with a strong
demand for wood, especially for house and ship building. In
addition most settlements were located in the upper reaches of
the mountain relief for defence reasons.
    The steeper parts of the typical Mediterranean mountain
relief, however, are especially endangered by erosion after
extensive clearing. The wide-spread Neogene rocks of the geologically young relief are hardly consolidated and, therefore,
particularly prone to erosion.
    Consequently, with extensive clearing and cereal cultiva-
in the hill and mountain landscapes of the Mediterranean there
began a process of ecological degradation which in some regions
already led to desertification two thousand years ago. However,
these processes must be differentiated regionally and in terms
of the severity of their effect. Not all results of Mediterranean desertification were irreversible, even though severe
limitations of land use ensued, that could only be repaired in
parts after several hundred years. An example for the great
efforts made in southern European countries to rehabilitate
degraded areas are the reforestation projects. But what price
had to be paid for this rehabilitation!

3. Morphodynamic Processes

    The most important morphological processes triggered by
desertification in the semi-arid countries of southern Europe
are soil erosion, development of skeletal soils in combination
with decreasing nutrient content and a general change of runoff
mechanisms on the surface and in the soil. The impediment or
even destruction of the ability of the vegetation cover to
regenerate also causes micro- and mesoclimatic changes, i.e.
increased evaporation resulting in induration and development
of crusts (duricrusts, caliches), and decreased infiltration of
rainfall into soils resulting in higher surface runoff.
    On the whole, the antique extensive clearing of land
caused increased floods that threatened valley settlements or

made them impossible. Increased sediment transport, including gravels, from the slopes to the valleys and piedmont plains also buried settlements. Antique Olympia is an example for this. In general, thick sediment layers were accumulated in most valleys and on most piedmont plains in historical times, where-as all cultivated slopes were seriously degraded, unless pro-tective measures (slope terracing for example) were introduced. Eolian activities, i.e. deflation - unlike in the north African Mediterranean areas - play a minor role as soil desiccation is less severe than in North Africa.

4. Ecological Change

In spite of differences in the degree of desertification in the countries of southern Europe, general characteristics of ecological change common to all can be observed. Intensified washing-away of soils from all insufficiently or not terraced slopes in the Mediterranean hill and mountain landscapes dis-placed these soils and their nutrient contents down the slopes onto their lower parts, into the valleys and onto the foreland plains. Consequently, the possibilities of cultivation without the employment of special techniques on the slopes deteriorated steadily during the course of history, whereas land use poten-tial in the valleys improved. This, however, could not be uti-lized before modern hydrotechnical measures made possible the flood control necessary for drainage and irrigation. This de-velopment essentially is one of the last hundred years. During this most recent period considerable changes in cultivation and agrarian production took place. An example are the products of the extensive irrigation areas (citrus fruits, vegetables, rice etc.) existing today. Fig. 1 shows the development schematically.

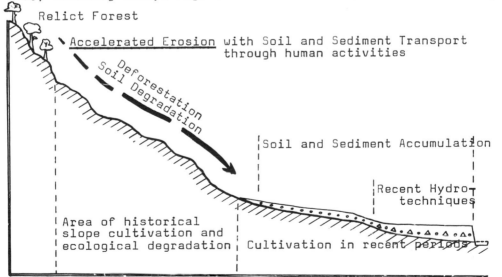

Fig. 1: Model of ecological change in Mediterranean landscapes in historical times.

Through the degradation of the natural ecological potential and the deterioration of soil quality on the slopes, land
use potential in the valleys and on the plains was improved.
Continuing forest clearing, however, steadily enhanced flood
dangers in the lowlands, so that the upgraded land use potential could not be taken full advantage of before adequate water
control measures were developed.

5. Combatting Desertification

Even in early historic times, the inhabitants and users of the
Mediterranean landscape were acquainted with and carried
through measures against soil erosion. This includes the widespread terracing of slopes, especially stone terraces and walls.
Even in more level areas earth walls were often constructed to
protect the soil cover.
    Wherever extensive land use, especially cereal cultivation,
was practiced in semi-arid to arid climates without sufficient
protective measures, gradual desertification followed in the
form of overall soil erosion and degradation. In wide areas
cultivation had to be given up. Most often they were converted
to grazing land, for permanent or or seasonal (transhumance)
use. This meant, however, that such areas could hardly regenerate ecologically and that nowhere a sound tree stock could
redevelop. At best, forms of secondary vegetation like Macchia,
Phrygana or Garrigue developed, one and all stages of degradation of the Mediterranean forest. In arid areas a form of
steppe or desert steppe with a very much degraded xerophytic
vegetation developed. Here deflation can be observed.
    Of paramount importance for combatting desertification are
all measures designed to control surface runoff. These measures
will have to focus on the construction of earth dams as they
have been known from many historical landscapes of the Mediterranean. In Greece and wide parts of Yugoslavia destruction
of the natural ecosystem led to considerably accelerated processes of karstification.
    In the last thirty years great efforts have been made in
the Mediterranean countries to combat ecological degradation
by reforestation. Especially in Spain the catchment areas of
the rivers draining into the Ebro Basin and wide expanses in
Castile have been reforested.
    All measures against ecological degradation and desertifi-
cation must, however, take into consideration the natural
degree of aridity in the various regions and will, therefore,
differ from one place to another. They must concentrate first
and foremost on the regenerative ability of the plant cover
and a better control of torrential surface runoff.

6. The Question of Climatic Change

    The question of climatic change in the Mediterranean in
historic times has been widely discussed. Especially archaeologists have often claimed that the decline of originally
widely cultivated areas since the Imperium Romanum has to be
attributed to a deterioration of climate. The question whether
this was a matter of a general depression of temperature or a
reduction of rainfall remained unresolved. Direct climatologi-

cal proof is not available.

Changes in water balance and surface runoff, however, strongly depend directly on the vegetation, not least on the existing forest stand. This brings us back to the consideration of deforestation since antique times and its consequences. From investigations carried out in Tunisia in 1976 we can deduce with more likelihood that the deterioration of large areas which eventually rendered them unsuitable for further cereal cultivation is due to a man-made degradation of the ecological potential. This includes micro-climatic changes caused by desertification with severe effects on the soil-water balance.

7. Examples from Western Mediterranean Countries

Spain comprises the most extensive part of the Mediterranean arid zone. Parts of the Ebro Basin, the north and south Meseta, as well as the Guadalquivir Basin are semi-arid with four arid months (Thornthwaite), the south-east coast with 6-8 arid months is fully arid. Desertification consequences are wide-spread in extended areas of grain cultivation. Denudation, accelerated sediment transport into the valleys and gully erosion in soft rocks are characteristic phenomena. In recent times attempts have been made to rehabilitate parts of the destroyed land through a system of slope terraces and terraced reforestation. These areas are to be integrated into better controlled land use taking into consideration the whole ecosystem. This is practiced especially around the reaches of larger water reservoirs ('pantanos').

In Sicily especially areas with less than 600 mm of rainfall and six arid months are prone to desertification. The advance of xerophytes has already seriously changed the natural character of the vegetation. After 2,500 years of intensive land use the natural vegetation in some parts can no longer regenerate and assert itself. This results in increasing occurrences of 'calanchi' and 'frane'. Slope soils are heavily degraded, soil erosion from the slopes to the valleys has reached a maximum. Sicily is a particularly good example for desertification ensuing from increased morphodynamic activity through torrential rains.

Controlled land use as a measure against historically advanced desertification is indispensable. Degraded slopes must be left unused for considerable time and restored by afforestation. Further extension of irrigated cultivation in the valleys can compensate this loss in area, but this will also have tenurial and social consequences. Especially in Sicily attempts at change in this area meet with considerable difficulties. Desertification problems must also be considered in the social and economic context, combatting it requires consideration of the socio-political level!

8. Conclusion

In Europe desertification is wide-spread only in the semi-arid reaches of the Mediterranean countries. In their semi-humid and humid parts it is observable only very locally, although degradation with serious damage affecting the ecosystem does otherwise exist in other forms. Different types of ecological degradation should be differentiated.

On the UNCOD map of desertification (United Nations Conference on Desertification, UNESCO-FAO, 1977) only a large part of the Iberian Peninsula is classified as prone to desertification hazards in Europe. This must be corrected inasmuch all countries with a semi-arid climate are endangered.

In all Mediterranean countries the process of soil degradation is the result of the very old cultivation history of the mountain relief. Protective measures are equally old (terraces), but have recently been seriously neglected in the course of increasing mechanization even on the slopes.

A better, more ecologically focused control of land use methods should, therefore, be a priority task in all countries of the European Community. To date this seems to play a minor role in the predominantly economic discussion.

REFERENCES

1. UN-Desertification: Its causes and consequences. Pergamon Press 1977.
2. BAKE, G. (1980). Physisch-geographische Grundlagen und Dynamik von Desertifikationserscheinungen in SW-Sizilien. Diss. Univ. Hamburg, 129 p.
3. MENSCHING, H. (1980). Desertification. Ein aktuelles geographisches Forschungsproblem. Geogr. Rundschau, 9, 350-355
4. MENSCHING, H. et al. (1976). Desertification im zentral-tunesischen Steppengebiet. Nachr. Akademie d. Wiss. Göttingen, No. 8, 1-20
5. MENSCHING, H. (1983). Die Verwüstung der Natur durch den Menschen in Historischer Zeit: Das Problem der Desertification. In: Natur und Geschichte (Ed. H. Markl). Oldenbourg Verlag München, 147-170

THE SCALE FACTOR IN RELATION TO THE PROCESSES INVOLVED IN

"DESERTIFICATION" IN EUROPE

A.T. GROVE
Director of Centre for African Studies
and Fellow of Downing College, Cambridge.

Summary

Desertification in Europe is more appropriately called environmental degradation. Such degradation could conceivably be the outcome of natural events, man's activities or interaction between the two. The processes involved and the appropriate responses to them can be considered in relation to the scale on which they operate.
a) On the global scale of $10^8 Km^2$ we have evidence of a warming in this century to which artificially induced increase of atmospheric CO_2 may have contributed. The research supported by the European Community might help to throw light on the nature of this phenomenon with particular regard to the environmental consequences for different regions within the Community.
b) On the sub-continental scale of $10^6 Km^2$ environmental impact involves the atmosphere and oceans. "Acid precipitation" with its effects on vegetation, rivers and lakes is on the scale of the Community. Pollution of coastal waters from ship and urban sources is on a similar scale and deserves Community research and preventative measures.
c) On the sub-national scale of $10^4 Km^2$, processes of environmental degradation include accelerated erosion and sedimentation, soil salinisation, falling water-tables, deterioration of the plant cover and urban sprawl. Although these processes operate locally, in individual river basins for example, it can be argued that they have a Community-wide dimension.
(i) The causes lie in changing socio-economic conditions within the Community.
(ii) The problems involved are similar in quite widely separated areas. Collaborative research could provide the economical and speedy way to obtain results applicable in more than one Community country.

Desertification is a term that is usually applied to those parts of the arid zone where biological productivity has been reduced as a result of man's use, or rather misuse of the land. It has not usually been applied to Europe, most of which lies outside the arid zone as it has come to be defined conventionally. The exception is Spain where Dregne (1) states that approximately 50 per cent is arid and that in the arid regions about 60 per cent is moderately desertified and 40 per cent severely desertified. Some other parts of southern Europe are similarly, though less severely afflicted; conditions in Turkey are believed to be more serious than in Spain. [See for instance, Thirgood (2)]. Elsewhere in Europe land is being degraded, but because the areas concerned are not arid and may even be unusually wet, the term desertification is unsuitable. I consider it should not be used in these circumstances. Land degradation or environmental degradation are more appropriate terms and I would suggest that we should use them in the future when referring to conditions in Europe outside the arid zone.

Environmental degradation can conceivably be the outcome of natural conditions. The kind of drought that has afflicted the lands bordering the southern Sahara for most of the last 16 years seems to me to be a good instance of this. The desert border has effectively moved south and would have done so without any increased pressure from human activity there. It seems that there has been a step down in precipitation for some reason as yet unexplained, which is without precedent in the existing records for the region. Such an unpredictable climatic event could conceivably affect Europe. Long-continued severe drought could cause great damage to the environment and could greatly affect both agricultural and industrial economies, especially in southern Europe as recent droughts of comparatively moderate magnitude have indicated. Probably the most significant climatic shift that has affected Europe in recent times is the warming that has been taking place globally since the middle of the last century and which has caused glaciers to retreat in some cases by as much as 2 kilometres. The effects of this warming have on the whole been beneficial, but a reversal could occur at any time, so far as we can tell, leading to a lowering of the snowline and, especially in the maritime upland areas of north-west Europe, a lowering of the altitudinal limits of cultivation through a hundred metres or more.

In both environmental degradation in Europe and desertification in the arid zone, man has more often been identified as the culprit than nature. If he is the cause then he has the opportunity to undo the damage he has done or at least prevent further deterioration. Whereas climate seems to fluctuate about a long term mean on the time scale of the last few millennia, man's pressure on the environment has been increasing at an accelerating rate for the last three centuries. In Europe it is the outcome not so much of population increase (that phase seems to be past), as of increasing energy use. Consumption of fossil fuels (and latterly of nuclear power) has allowed increased mobility of people and goods, the concentration of settlement in urban areas, intensification of agricultural production, and more extensive recreational activities and tourism; these seem to be the increasingly active ingredients in the environment.

We can look upon environmental degradation as being the result of the interaction of two systems, the natural physical and biological system on the one side and the human economic and social system on the other. Whereas the natural system tends to oscillate over a period of centuries, the human system becomes more energetic and effective from year to year. The human system can respond to extreme natural events involving say volcanism or drought by a spatial shift in activities so that dangers of breakdown are confined to small areas. Man can improve on nature as many agricultural landscapes demonstrate. But there are increasing signs that natural systems can suffer persistent deterioration that reduces their usefulness and attractiveness to man. Effective environmental action involves gaining a better understanding of the interaction of natural and human systems with a view to regulating both to their mutual longterm advantage. By longterm I mean over a century or so; one or two humans lifespans.

Spatially we have to concern ourselves with conditions on a number of different scales: a. global, b. transnational or sub-continental and c. sub-national to local are the ones I would be inclined to choose, with magnitudes of 10^8, 10^6 and $10^4 Km^2$. The European Community falls into the second of these categories and it may be worth considering how this scale factor affects the kind of research sponsored by the EEC that would accordingly seem appropriate.

a. One of the main concerns with the global environment at the present day is of course CO_2 loading of the atmosphere. All advanced countries are involved in research on this topic because all will be affected by it, though their vulnerability varies, and because it would seem likely that an effective response would lie in the hands of the industrialised coutries.

European Community research activity might be directed towards predicting the likely impact on its member states and possibly to allocating responsibility for mitigating harmful activities. So far the research that has been done serves to emphasise some important points that are probably of more general application. The recognition of an environmental hazard has emphasised our ignorance of some of the basic natural cycles and situations that underlie the immediate problem. Research in several coutries and in numerous scientific institutes, generated by environmental concern, has provided us with quite a new appreciation of the operation of the natural systems involved. In particular, it has become apparent that the natural level of CO_2 in the atmosphere is not constant and that it probably depends largely on the thermal structure and changing circulation patterns of the ocean and also on the activities of micro-organisms at the sea surface and macro-organisms on land. At the same time it has become apparent that if the CO_2 "hazard" is to be properly assessed and effective -ly handled, more research is needed, much of which at first might not have been seen as relevant to the original environmental problem.

b. On the transnational scale, a major environmental concern involves again primarily the atmosphere and its micro-components. "Acid precipitat- ion", like accentuated CO_2 loading, appears to be the result of the burning of fossil fuel, with pollutants from some parts of the European Community, for example, affecting unfavourably the soil, forest and water conditions in other parts. Again, research is immediately directed towards identify- ing the nature of the hazard with a view towards taking palliative measures in the areas being degraded and preventative measures in the source areas of the pollutants. At the same time it is becoming clear that research is required into aerosols more generally, of natural as well as of artificial origin, into soil processes and the whole question of the changes that are taking place in soils and peat bogs, lakes and rivers on account of other pollutants of industrial origin. New exploratory tools and techniques, involving for example the magnetic properties of sediments, have been designed to trace the sequence of processes that have been operating in the landscape through historic, prehistoric and geologic time.

c. The majority of the processes involved in desertification and land degradation in Europe appear at first sight to be operating on the provincial or local scale. Accelerated erosion and sedimentation are commonly the outcome of particular land use practices operating in a particular locality, say with steep slopes, erodible soils and occasional intense rains. Urban sprawl is often concentrated in certain coastal localities, for instance where flat sandy areas are conveniently located with access to airports. Salinisation of soils takes place where ground- water reaches to within 2 or 3 metres of the surface as a result of irrigation water leaking from conduits or being applied too freely. Water shortages and pollution develop where competition for limited groundwater supplies results in overpumping and migration inland of seawater. The problems vary from one locality to another; the physical causes are local and it would appear that remedial measures can be taken by the local community, possibly with assistance from the state. Why should there be any need for Community involvement with environmental degradation of this kind? I am not sure that I have the right answer to this question. It

seems to be one worth asking and attempting to answer in the context of
this meeting.

I presume that the objectives and principles of the Environmental
Action Programmes include the rational management of natural resources,
with the aims of protecting and improving the environment. I understand
that the Community's environmental policy is effectively constituted by
the legislation it makes in the form of Directives (3). Much of the
Community's environmental legislation in fact comes within the context of
the Common Agricultural Policy, in the form of Directives concerned with
countryside protection in agriculturally less-favoured areas, primarily
intended to ensure the continuation of farming, thereby maintaining a
minimum population level, in areas that are mountainous or otherwise
infertile and relatively impoverished. The effects on the environment in
the form of numbers of people who continue to live in such areas, the
livestock they own and the forest they plant are not to be ignored.

My own understanding of the possible benefits from the involvement of
the Community in supporting an environmental research programme relating
to desertification in Europe are twofold. Firstly, it would seem that in
many instances the basic causes of environmental degradation on the sub-
national, provincial or local scale lie in socio-economic conditions
generated in the Community as a whole. Secondly, the processes involved
often have a common base in several member states so that it might be
expected that collaborative research may turn out to be more economical
and speedy than individual uncoordinated research projects.

(i) Amongst the socio-economic conditions on a Community scale that
I have in mind is the expansion of tourism. Tourist receipts are very
large, running into tens of billions of dollars annually (e.g. Spain $6.3
billion in 1981) and constituting a large proportion of "exports" (about
20 per cent in Spain and Greece). The number of overnight stays by international tourists in Italy numbered 84.4 million in 1981. In these
instances, most of the tourists come from other Community countries. As
working people procure greater paid holiday entitlements so the demand
increases. The majority of the tourists concentrate in relatively small
areas, especially major coastal resorts, and seasonal peaks facilities are
overloaded.

Research has identified links between tourism and environmental
degradation (4 & 5). It is evident that, locally, more extreme conservation measures are being required. Solutions to the problems arising require
the assessment of tourist carrying capacities, restrictions on tourist
development in areas at risk, the flattening of demand peaks by staggering
holidays and various marketing measures. Eventually overall spatial
planning that incorporates the tourist element will probably be needed.
Such measures, involving costs to local governments, may require financing
from Community sources.

Other socio-economic trends having environmental consequences include
the abandonment of traditional seasonal mobility, especially pastoralist
transhumance with summer occupance of upland grazing areas. This system,
typical of southern Europe until a few decades ago, has declined as a
result of the attractions of increased employment opportunities in the
towns. The effects on rural communities, landscapes and land management
deserve attention throughout the regions concerned.

(ii) In order to predict future environmental change, research of two
kinds is needed: A, into the current situation and the processes operating
in the landscape at the present and, B, into the development of the
present landscape through time. In both cases some kind of sampling
programme would seem to be required. I have suggested that this might

involve a kind of nested sampling, with regional studies increasing in intensity from the national level, through river basins as units, to valley transects within those basins, and particular sites within those transects.

A. The intention would be to establish "baselines" from which changes in the future could be measured; the range of phenomena involved varying according to the scale of the surveys. While national surveys might utilise existing statistical data on land use etc., remote sensing methods might be employed for the river basin studies and intensive ground observations for the more detailed surveys. Ideally such work would be centred at existing research stations within the selected river basins and one might look towards the standardising procedures throughout the Community countries involved with exchange of personnel between countries and possibly the involvement of University staff and students. The need for such studies would seem to be greatest in places where pressure from tourist and other development is beginning to increase rapidly.

B. The results of studies of present conditions become more meaning-ful in the light of knowledge of the history of the landscape development. Palaeoenvironmental studies are becoming more sophisticated and productive as new techniques are employed. Already much information has been gained, for example in connection with the Santorini (Thera) explosion about 1450 BC, (6). Europe offers particularly favourable opportunities for such research because of the abundance of archaeological sites and the expertise available. It would be advantageous to arrange for the studies of current processes to be coordinated with those of the palaeoenvironment and if possible to be conducted in the same areas.

A subject of particular interest is the origin of the valley floor sediments and coastal aggradation in the Mediterranean basin within the last two millennia. The Younger Fill, as it has been called, that forms the terrace alongside most stream courses is believed to have accumulated in post-Roman times and to have been dissected a few centuries ago (7). Questions arise as to whether it represents a climatic fluctuation or several storm events and what part was played in its formation by human activity.

A recent geo-archaeological study of medieval Alzira in southeast Spain exemplifies the kind of investigation that can be made to throw more light on the subject (8). From the results it appears that a long-term shift to high-energy floods in the region around Valencia about the 11th century is to be explained by watershed deforestation; the aggrading floodplain of the Jucar River became increasingly prone to flooding on account of a fan deposited by the catastrophic flooding of a tributary downstream of Alzira in 1517; between 1300 and the early 18th century, a marked clustering of destructive floods was associated with aberrant climatic patterns. Such, in summary form, is the background for the current behaviour of the Jucar river.

Studies of this kind involve the collaboration of archaeologists and historians as well as earth scientists. In England, the Domesday survey of 1086 provides a data base of great value against which subsequent changes in land use can be measured. It seems that similar surveys on a smaller scale were made by the Normans in Sicily. With the accumulation of comparable records from elsewhere and the addition of data from geomagne-tic and other studies of core material from lake beds and coastal sediments, it should be possible to sharpen considerably the picture we have of the evolution of the Mediterranean environment.

Europe has long supported a large and growing agricultural population and has been one of the most productive regions on earth. The population

of the European Community now compares with that of the USSR or the USA
and the Community's agricultural output and its industrial production are
also comparable with those of the two great sub-continental states. Yet all
this activity is concentrated in a fraction of the area of the USA or USSR.
The risks of environmental degradation ·in such circumstances are evident
and great care must be taken to conserve the land and water resources on
which so much depends.

Perhaps the time has come for a new Domesday, a European Domesday,
an assessment of the landscape in preparation for the new century so near
at hand.

REFERENCES

1. DREGNE, H. (1983). Desertification of Arid Lands. Chur, Switzerland:
    Harwood Academic Publishers.
2. THIRGOOD, J.V. (1981). Man and the Mediterranean Forest. London:
    Academic Press.
3. HAIGH, N. (1984). EEC Environmental Policy and Britain. Environmental
    Data Services, London.
4. O.E.C.D. (1980). The Impact of Tourism on the Environment. Paris.
5. MATHIESON, A. & WALL, G. (1982). Tourism: economic, physical and
    social impacts. London: Longman.
6. DOUMAS, C. (ed.) (1978 & 1980) Thera and the Aegean World, Volumes I
    & II. London: 105-9 Bishopgate.
7. VITA-FINZI, C. (1969). Mediterranean Valleys: Geological Change in
    Historical Time. Cambridge University Press.
8. BUTZER, K.W., MIRAILLES, I. & MATEU, J.F. (1983). Urban Geo-archaeol-
    ogy in Medieval Alzira (Prov. Valencia, Spain). Journal of Archaeology
    10: 333-349.

DESERTIFICATION IN A CHANGING CLIMATE
With a particular attention to the Mediterranean countries

A. BERGER
Institut d'Astronomie et de Géophysique G. Lemaître
Université Catholique de Louvain
2 Chemin du Cyclotron
1348 Louvain-la-Neuve

Abstract

There is a duality between climate and general circulation. Southern
Europe belongs at present to the dry summer Mediterranean climatic
zone, situated between the woodland suboceanic climate of the cool-
temperate zone, (which is related to the transient eddies associated
to the polar front), and the dry climates of the high pressure subtropical belt. Any change in the general circulation resulting, for
example, from changes in the boundary conditions which characterized
it, will lead to a change of the overall climatic pattern. Such
changes, as occured 125,000 YBP (peak of last interglacial), 18,000
YBP (maximum glacial advance of the Wurm),9,000 YBP (peak of Holocene
interglacial) and as expected from a doubling of the atmospheric CO_2
in the mid-21th century, will be reviewed in order to stress the
potential subsequent change in the vulnerability of the environment
to desertification in these regions.

1. THE MEDITERRANEAN CLIMATES AND THE GENERAL CIRCULATION OF THE ATMOSPHERE

The term general circulation of the atmosphere is commonly used to
refer to the more or less permanent global wind systems of the troposphere
and lower stratosphere. It is customary to describe these motions in
terms of long time average states and supplement these with considerations
of the general aspects of the disturbances superimposed upon them. As a
matter of fact, this dynamics of the atmosphere and its interactions with
the other parts of the climate system (hydrosphere, cryosphere, litho-
sphere, biosphere) are directly related to the world global climates.
     Although a cursory inspection of a climatological chart suffices to
show that the influences due to oceans and continents, mountain ranges ...
(differentiation in longitudes) are very large, zonal patterns are clearly
present. Consequently, to obtain an idea of the principal characteristics
of the general circulation, and thus of the climates, we might consider
zonality and seasonality first.
     As both the maintenance of a balance budget of heat and the conserva-
tion of the total angular momentum of the earth-atmosphere system are im-
posed upon the circulation, the zonal winds near the earth's surface and

the corresponding meridional motions are best described by the three-cell scheme (Lorenz, 1967).The gross features of that atmospheric general circulation are :

(i) the Hadley circulation in the tropics where the main transports occur through a meridional circulation,

(ii) the subtropical high pressure belt (STHPB) where the divergent wind system and the associated large scale subsidence lead to rainfall which is both meager in annual total and scanty in all seasons (characteristic of the earth's great deserts),

(iii) the sloping convection at mid and high latitudes,

(iv) the large scale eddies which are the dominant means of transport of heat and momentum in mid latitudes,

(v) a polar front associated with the polar front jet,

(vi) the subtropical jet associated with break in tropopause at $\sim 30°$ latitude.

In most regions, the key to climate lies in the behaviour and position of these characteristic features of GCA and their seasonal change. For example, the poleward margin of the STHPB is covered by high pressure in the summer season only but by circumpolar westerlies with their travelling disturbances in the winter. This is precisely the regime with winter rainfall or mediterranean type conditions. The geographical position of the Mediterranean Basin, at the border between two climatic zones, makes that region potentially highly vulnerable in a dynamical climatic system : slight variations in the general climate pattern may induce there strong local changes.

In practice the longitudinal symmetrical structure is complicated in two ways (e.g., Hare, 1983). The first is that the STHPB is markedly cellular at low altitudes and the second is the development of the great monsoonal anomalies of Africa, Asia and to a minor extent North America. These separate cells permit the systematic migration of equatorial airstreams polewards round the western ends of the cells, and of mid-latitude airstreams equatorwards round the eastern ends. In northern hemisphere, these cells tends to remain anchored over the ocean thus providing certain preferred longitudes for persistant poleward or equatorward airstream movement.

Thus in the summer season, the persistence of high pressure between the Azores and Bermuda produces almost continuous northerly or north-easterly airstream over the western most Sahara, cool and moist relative to the air over the interior. North and west flows characterize also southern Europe. All these currents are strongly diffluent and divergent, and hence encourage subsidence. All have the effect of intensifying the aridity but of mitigating the summer heat.

In winter, the equatorial displacement of the Azores-Bermuda High allows secondary storm tracks to cross the Mediterranean region.

All the climatic regimes have unsually high internal variability in rainfall and cloudiness and it is important to distinguish between interseasonal and interannual changes. The causes of these variations from year to year, as for the interseasonal changes, are mainly dynamical. There may for example, be a persistant shift in the latitude of STHPB and of the airstreams that flank it (the world-wide precipitation anomalies of 1970-72 is a good example (Lamb, 1977)). Other rainfall anomalies can be traced, however, to changes in the east-west cellular structure of the

STHPB and hence to altered airstream movement (e.g. inland Australia excessive rainfalls over 1972-75 (Hare, 1983)).

The longitudinal contrast among the Mediterranean climates is thus important to consider. From a principal component analysis and a cluster analysis, Goossens (1985) was indeed able to differentiate 5 sub-Mediterranean climatic regions ranging from west to east with decreasing precipitation.

In summary, to quote Hare (1983) : "Both the characteristic distribution of the continental rainfall belts and the anomalies to which the dry climates are subject, are controlled mainly by the atmospheric general circulation and, only to a minor extent, by local circumstances. It follows that the root causes of drought or excessive rainfall are largely global and that the problem of predicting them is also global".

The review presented here is based on a large number of papers which can be found in specialized journals or books (e.g. Berger, 1981; National Research Council, 1982), but we will refer mainly to the comprehensive works of Prof. H. Flohn (1980), Prof. K. Hare (1983), Dr. T. Wigley and his team (1980-1984), and of the U.K. Meteorological Office.

2. GENERAL CIRCULATION MODELS OF CLIMATE

In seeking possible causes of major climatic fluctuations, one may contemplate changes in the external forces acting on the atmosphere, internal changes within the atmosphere-ocean system itself or, more likely, an interactive combination of both (including anthropogenic forcing).

The global climate could thus be modified by changes in :

1) the incoming solar radiation
2) the atmospheric composition, e.g. CO_2, dust, ozone, water vapour content
3) the cloudiness
4) the albedo of the Earth's surface through changes in soil moisture and vegetation, ice and snow cover
5) the oceanic circulation leading to changes in sea surface temperature and in the transport of heat and moisture.

The boundary conditions associated with these changes can be used to study climate sensitivity with climate general circulation models (GCM) that represent the global circulation in terms of the mathematical equations governing large scale atmospheric motion. GCM's do not address the problem of long-term climatic change directly. Rather they attempt to explain in dynamic terms the complex of processes which, by balancing the earth's radiation budgets, maintains a particular climate in equilibrium.

The purpose of this paper is to review some of these experiments made with GCM's in order to investigate the climates which are in equilibrium with particular boundary conditions revealed by the history of past climates or assumed to be the consequences of man's activities on our future climate (Table 1). This analysis will be done in order to stress the potential subsequent changes in the vulnerability of the environment to desertification particularly in southern European countries.

Table 1. Particular climatic situations analysed in order to assess the vulnerability of southern European countries to desertification.

---

1. Past climates
   Tertiary Climate
   Eemian Interglacial (125,000 YBP)
   Last Glacial Maximum (18,000 YBP)
   Deglaciation (Wurm - Holocene)
   Holocene Optimum Climate (9,000 YBP

2. External forcing changes
   Solar Constant
   Milankovitch Theory

3. Internal Changes
   Sea surface temperature
                anomalies
   Soil moisture experiment

4. Anthropogenic Changes
   Charney effect
   Increasing carbon dioxide

These results of first generation climate sensitivity experiments are, of course, open to question. The models incorporate only some of the physical processes of the climate system and these processes no doubt are imperfect. Also the observational basis for the assumed parameter changes (boundary conditions) is not necessarily adequate. Lastly, there remains the difficult problem of model validation. These deficiencies are recognized and increased sophistication in the experimental design of future climate sensitivity studies are in progress.

3. LATE TERTIARY UNIPOLAR CLIMATIC ASYMMETRY

In his investigation of possible climatic consequences of a man-made global warming, Flohn (1980) was particularly interested by the climate of the Late Tertiary, a time with an open Arctic Ocean and a glaciated Antarctic Continent. The evidence of such a coexistence accumulated during the Deep Sea Drilling Project are summarized from his paper.

This highest Antarctic glacial maximum was reached at the end of the Late Miocene (or Messinian 6.5-5 Myr ago - Myr = 10^6 yr) with the ice volume 50% larger than it is to-day. During this remarkable cooling period, the Arctic Ocean had a cool to temperate climate. But one of the most important consequences was a glacial-eustatic drop of the sea-level of 40-50 m below today's level. During this drop, the Gibraltar Strait (or its predecessor), fell dry isolating the Mediterranean, which evaporated completely 8-10 times to a depth of 3,700 m and filled again, indicating a quasi-cyclic behaviour at a time scale of about 10^3 years, which lasted for about 0.5 Myr, and could be caused by changes in the orbital elements similar to those effective during the Pleistocene (Berger et al., 1984). This multiple desiccation of the Mediterranean aggravated the arid conditions in southern Europe. Even south-central Europe was partly arid, with steppe or desert vegetation near Vienna, and SW Germany was also drier than it is now.

The whole sequence of the displacement of northern boundary of the NH arid zone has been compiled from maps of the position of the evaporite belt during the last 50 Myr. It indicates a southward shift from an average latitude of about 47°N in the Early and Mid Tertiary (30-50 Myr ago) to 42°N in the Miocene/Pliocene (5-15 Myr ago), and to 38°N in the Pleistocene.

In such a unipolar climatic asymmetry, the average boundary between the Hadley and the Ferrel cells has been extrapolated using Flohn's index of climatic zonality.

From such a simple extrapolation based on the temperature distribution above an ice-free Arctic, the shift would reach in summer probably not more than 100 or 200 km, but in winter the subtropical belt may be displaced by some 800 km or even farther north. This would drastically reduce the extension of the subtropical belt of winter rains, now supplying California, the Mediterranean, the Middle East up to Turkestan and the Panjab. The belt 35-50°N would probably be frequently affected by summer droughts.

One of the serious threats might be desiccation of the Mediterranean countries especially south of about latitude 42°N, but the frequency of summertime droughts should increase even well north of this latitude.

It must be pointed out that any rational attemp at predicting the effects of disappearing Arctic Sea ice (due to CO_2 increase for example) remains very difficult. Indeed, since the late Tertiary the altitude of the mountain blocks has increased and those (particularly north of the Himalayas) largely control the planetary circulation patterns. However, a rough estimate of the global water budget based on the displacement of the northern subtropical anticyclone to latitude 35-45°N support a general increase in aridity, with a loss of water on the order of 4-5% at least.

This Arctic ice cap was also assumed to melt and was replaced by a water surface held at 0°C in an experiment with the U.K. Meteorological Office 5 level model (Newson, 1973). Among the major unexpected results, this produces a significant cooling of up to 8°C in variable latitudes over western Europe and an increase of 2°C over the Mediterranean basin.

4. EEMIAN LAST INTERGLACIAL

Recent investigations of ocean cores, as well as of continental loess deposits, indicate that over the past 1 million years a sequence of 10 large-scale glaciations of northern continents has occured (at least 17 over the past 2-2.5 Myr), interrupted by the same number of interglacials with a climate similar to the present (Berger, 1979a, 1981). The last interglacial at around 125,000 YBP lasted roughly 10,000 years and was probably the warmest of all with a sea level higher than today's by 5-7 m.

In northern and eastern Europe, the climate was much more oceanic than it is at present (temperatures were 2-3°C higher than today and slightly more humid; hippopotemuses, forest elephants and lions roamed southern England!). This holds true for the Mediterranean area also, where data indicate January temperatures 3 to 4° higher, July temperatures about 2°C higher, and a perceptibly higher total rainfall. These conditions are also valid for the penultimate warm period (Holstein, centered $\sim$210,000 YBP). It is, however, not clear whether, as in the Holsteinian warm period, the principal rainfall occured during the summer months. Nevertheless, data show that the water supply for the plant community during the Eemian Climatic Optimum appears to have been more favorable than the present one (Frenzel, 1973).

This summer situation was simulated with the EERM Spectral General Circulation Model (Royer and Pestiaux, 1984) where the lower boundary conditions were kept to their present values but where insolation was changed and set to its 125,000 YBP value (11% higher than now; Berger, 1978,

1979b). During this simulated interglacial mode, the latitudinal temperature distribution indicates a temperature increase of more than 3°C between 50 and 70°N (the average over NH amounts 1.3°C). Precipitation increased by 22% globally over the Earth. More interestingly for our purpose, the surface temperature increased in Europe by 1°C and in the Near East decreased by a few tens of a degree. The difference between precipitation and evaporation increased over the northern hemisphere continents; as the reservoir has only slightly increased in Greece, runoff must have been more intense there.

This warm phase was interrupted 115,000 YBP ago by a cold episode particularly apparent in continental records (Woillard, 1979). In order to understand the physical mechanism of such abortive glaciation, two complete annual cycles of climate were performed with the same model (Royer et al., 1983) for the insolation corresponding to 125,000 YBP and 115,000 YBP (a drop of more than 60 W m^{-2} - 15% of present value - in all northern latitudes during June–July and more than 80 W m^{-2} in southern high latitudes in September and October (Berger, 1983)). Comparison of the statistics of the two experiments shows an annual mean cooling of more than 2°C over Canada with increased precipitations for the 115,000 YBP case, which could act as a trigger for the initiation of an abortive Laurentide ice sheet (in this case, cold summers are not compensated by warm winters in high NH latitudes although the Earth is in aphelion in July). The map of surface temperature differences shows also a colder region extending from the Mediterranean to Siberia with a decrease in the soil water content over southern Europe. In sea level pressure, more cyclonic conditions prevail in January and more anticyclonic in July, which might reflect a greater aridity in summer there.

5. THE LAST GLACIAL MAXIMUM 18,000 YBP

5.1. Paleoclimatic reconstruction

In the Northern Hemisphere, the 18,000 YBP world differed strikingly from the present in the huge land-based ice sheets, reaching approximately 3 km in thickness, and in a dramatic increase in the extent of pack-ice and marine-based ice sheets (CLIMAP[*], 1976). In the Southern Hemisphere, the most striking contrast was the greater extent of sea-ice. On land, grasslands, steppes and deserts spread at the expense of forests. Sea level was lower by at least 85m but sea-surface temperature anomaly averaged only -2.3°C over the entire ocean. In the Northern Hemisphere, the large shift in the path of Gulf Stream waters caused the Atlantic to have the highest anomaly. In the Southern Hemisphere the marked cooling along the South Equatorial Current gave the Pacific the largest average difference in temperature.

Additional insights into the spatial pattern of climate change can be gained from litterature (July mean temperature in NE and Central Europe was 7-8°C lower) and here we will focus only on the Mediterranean region. The area of pronounced change in the North Atlantic extended southward along Spain and included the northwest Mediterranean. The eastern Mediterranean however, experienced a temperature decrease of only 1° to

---

[*]CLIMAP is an acronym for Climate, Long Range Investigation Mapping and Prediction.

2°C (in southern France, up to 10°C). Thus, whereas the temperature difference between the Balearic Islands and Crete is only 1° to 2°C today,
the corresponding east-west gradient during the ice age was probably as
large as 10°C.

5.2. Saltzman-Vernekar experiment (1975)

A model governing the zonally averaged macroclimate applied to the
present surface boundary conditions provides us with an acceptable first
approximation to the observed climatic zonation. Its equilibrium solution
in response to the glacial surface boundary conditions shows the glacial
atmosphere to be colder and drier than at present, with an intensified
polar front, stronger mean zonal and poloidal winds, more intense transient baroclinic eddies (storms) transporting heat, momentum and water
vapor polewards at higher rates, reduced precipitation and evaporation,
and an equatorward shift of the climatic zones particularly at higher latitudes. The zonal mean surface and mid tropospheric temperatures are uni-
formely cooler than at present and shows an enhanced low level polar front
near the edge of the ice sheet at 55°N in winter. Associated with this
winterpolar frontal zone are southward shifted surface polar easterlies
and midlatitude westerlies. The boundary between the midlatitude surface
westerlies and the trades in winter is essentially unchanged however, resulting in a contraction of the surface westerly zone. In summer, there is
a southward shift of roughly 4 degrees of latitude in both the polar easterlies and westerlies.

The hydrological cycle is characterized by a global decrease in both
evaporation and precipitation along with an increase in the poleward eddy
flux of water vapor. The climatologically significant net effect is a
somewhat equatorward-shifted pattern of (E-P) with increased amplitude in
winter, resulting in a more positive water balance from 25 to 75°N and a
less positive balance in the remaining tropical and polar latitudes.

5.3. Gates experiment (1976)

This was the first experiment to simulate global climate from CLIMAP
reconstruction of the 18,000 YBP climate conditions.

Perhaps of greater climatic interest are the simulated differences in
the surface air temperature for the ice-age and present July. These show a
marked lower ice-age surface air temperature immediately to the south of
the ice sheets with cooling of as much as 10° to 15°C over extensive
areas.

Accompanying these ice-age temperature changes are a number of systematic changes in the large scale tropospheric circulation. Prominent high-
pressure regions are simulated over the major ice sheets in the Northern
Hemisphere indicating a strengthened surface air flow from the east and
the northeast in the regions immediately south of the ice sheets, namely
in central Europe. Over western Europe, this is in sharp contrast to the
generally southerly and south-westerly flow infered from the results for
present July conditions. The westerly wind maximum simulated near 60°N
under present conditions is shifted to near 40°N and somewhat increased in
strength in the ice age, as a direct result of the southward displacement
of the zone of maximum north-south temperature gradient introduced by the
expanded ice sheets.

Along with the global ice-age cooling and circulation shifts, the simulated ice age global evaporation and precipitation are also reduced by
about 15 percent. The simulated distribution of precipitation shows a

notable reduction over the Scandinavian and Laurentide ice sheets and a reduction of the monsoonal rainfall over Southeast Asia. Associated with the southward shift to about 10°S of the Indian monsoon cell, is an increase of the precipitation along the eastern coast of Central Africa, extending over the Mediterranean (such features of the precipitation difference field are not evident in other simulations like Manabe and Hahn, 1977). Evaporation was less in this region during the ice-age. It must be stressed here that during the peak of the last cold period, stepps (not woodlands) were the characteristic elements in the northern Mediterranean indicating that precipitation was considerably lower (Frenzel, 1973).

Reduction in precipitation rate does not necessarily imply a change in the aridity of the continental surface. Therefore the geographical distribution of net water supply was computed. This quantity indicates the difference between the water gain at the surface (rainfall and snow melt) and the water loss (evaporation). Positive values mean increased availability of water for runoff or for soil moisture in continental regions and reduced salinity in oceanic regions. Such a detailed computation can be found in Manabe and Hahn (1977), whereas only zonally averaged latent heat balance of the atmosphere is available in Gates (1976).

5.4. Manabe and Hahn experiment (1977)

Manabe and Hahn (1977) conducted a similar integration of a global model developed at the Geophysical Fluid Dynamics Laboratory of NOAA. They found that the July climate of continental portions of the tropics in the ice age simulation is much drier than that of the modern climate, although there is relatively small differences in the amount of water available for runoff and soil moisture at most latitudes. Further analysis of the relative influence of sea surface temperatures and albedo show that the effect of increased continental albedo is mainly responsible for the weak Asian monsoon in the ice age simulation.

As one might expect, in tropical Africa, the rate of runoff of the ice age experiment was significantly less than that of the standard experiment, which is indicated by much geological evidence (Street and Grove, 1976).

5.5. Williams et al. experiment (1974)

In an other simulation of ice-age July climate carried out at the National Center for Atmospheric Research (Williams et al., 1974) with a five-level atmospheric model, the simulated sea-level pressure distribution showed more drastic changes than those reported by Gates (1976). For example, cyclonic activity was displaced southward in January and July by land and sea ice with the main activity south of the Laurentide ice sheet being in winter.

6. HOLOCENE WARM PERIOD

6.1. Paleoclimatic data

The history of the retreat of the North America and northern Europe ice sheets after the last ice-age maximum 18,000 years ago is fairly well known (Duplessy and Ruddiman, 1984). At the end of the warm Allerod (∿10,800 YBP) the tropical oceans became slightly warmer than now, which was most probably accompanied by a substantial weakening of the tropical Hadley circulation. Increasing evaporation in the tropics, lead to a rapid expansion of the tropical rainbelt towards higher latitudes. While the smaller Scandinavian ice sheet finally disappeared before 8,000 BP, the North America ice sheet persisted much longer (till about 5,000 YBP), creating a marked longitudinal asymmetry of the atmospheric circulation namely between 8,000 and 6,500 YBP. During this episode, Eurasia and Africa experienced the warmest epoch of the last 80,000 years but eastern North America remained relatively cool, certainly in summer, with frequent outbreaks of polar air. This caused a predominance of SW winds over the Atlantic. In winter, the frequent occurence of an anticyclonic ridge at $\sim$ 10°W would be consistent with frequent outbreaks of polar air also over central and eastern Europe, extending with copious precipitation into the Mediterranean and northern Africa.

At the peak of the postglacial climatic optimum ($\sim$6,000 YBP) temperature must have been roughly 2°C higher in midlatitude oceanic coasts, and a little bit lower in continental areas. In subtropical latitudes, the present arid areas enjoyed wetter conditions (humid period in the Sahara, as well as in the deserts of the Middle East up to Ragasthan). Evidence based on actual data supports the idea of a tendency towards synchronous shrinking of the arid belt at both flanks (Petit Maire, 1984).

For this Holocene warm climate serving as an analogue for our future climate, it must be stressed that two boundary conditions were met 6,000 YBP : the presence of limited but not negligible permanent ice sheets in eastern Canada and the lack of man-triggered desertification processes. The first one was definitely at the origin of increasing cyclonic activity in the Mediterranean in late spring, summer and fall, whereas the progressively increasing man-triggered desertification after 6,000 YBP has probably contributed to the slow and gradual desiccation process.

6.2. Adem et al. experiment (1984)

In order to better understand this deglaciation, a thermodynamic model was forced for 18, 13, 10, 8 and 7,000 YBP summers with the related radiation data, ice sheet boundaries and sea surface temperature. It showed that the insolation effect increases the global temperature of the NH by about 1°C at 10 and 8,000 YBP and about 0.5°C at 7,000 YBP. The effect is much more important in the continental areas reaching values as large as 3°C in North Africa. Between 10 and 7,000 YBP, the Mediterranean basin experienced a summer temperature 0.5 to 1°C higher than today.

6.3. U.K. Met. Office experiment (Mason, 1978)

Additional support of the astronomical theory of paleoclimates comes from an experiment with the 5-level Meteorological Office model in which the effects of variations in the Earth's orbital parameters on global temperatures around the northern summer solstice were assessed by running two integrations (Mitchell, 1976, 1977), one representing the solar radiation for the present day and the other for conditions prevailing 10,000 years ago when the Earth received about 7% more solar radiation in June than at present (Berger, 1978). The computed differences in zonal temperatures for June show that the troposphere was warmer everywhere 10,000 YBP, with a 6°C higher in the Arctic basin and 4°C at 30°N.

In fact, two kinds of model were used, one with fixed sea surface temperature and one with an interactive ocean. Whereas one might consider that the former would exaggerate the response of the atmosphere to changes in solar radiation, the latter should set a lower limit to the changes that one would expect in the real atmosphere. In the interactive experiment, global rainfall, 10,000 YBP, increased by 12%. Although there are some particularities in the control experiment, a comparison between the simulated July rainfall patterns for the fixed SST show an well marked increase of precipitation in the Mediterranean region.

6.4. Kutzbach experiments (1984)

Results similar to Adem et al. (1984) for 18, 9 and 6,000 YBP were obtained by Kutzbach and Guetter (1984). For July, the increased solar radiation at 6 and 9,000 YBP is associated with higher NH land temperature (1.5 and 1.8°C respectively) and increased monsoon rains (about 20%). For January, the decreased solar radiation at 6 and 9,000 YBP is associated with lower temperatures, whereas precipitation changes are small.

A more detailed analysis of the 9,000 YBP experiment shows that the amplified seasonal cycle of solar radiation in the northern hemisphere (e.g., in mid-latitudes 8% more in summer and 9% less in winter (Berger, 1983)) produced an increased seasonal temperature contrast with warmer NH summers and colder NH winters, and increased summer monsoon rains. This is also seen in southern Europe were temperature increased by 2°C in July and was more or less the same as today in January. In response to the increased heating in July, the decrease in sea level pressure amounts 4 mb in the Mediterranean region. In January, there is a general rise of pressure over cold land NH surfaces, but the pressure fall over the relatively warm Northern Atlantic Ocean extends well in western and southern Europe. This means that the cyclonic activity must have increased over the Mediterranean region particularly in summer, but also in winter. This is reflected in the zonal means, where for the monsoon region covering North Africa - South Asia, precipitation increased annually by about 250 mm (20% over the present value) and precipitation-minus-evaporation by about 90 mm (about 80% more than to-day).

7. MEDIEVAL WARMING

There is a collection of evidence which clearly shows that the Early Middle Age, the Medieval warm Period (750-1150 A.D.), was the warmest period of the last millenium (Lamb, 1977; Hammer et al., 1978; Wigley, 1977). The climate in high latitudes was unusually warm (temperature increase of around +1°C). Frequent droughts occured all over Europe south of 60°N. In the Mediterranean region and around, the Caspian Sea level was 32m lower than today, the Dead Sea was low, but the northern part of the Sahara was definitely wetter (Nicholson and Flohn, 1980).

One interpretation of these data suggests a northward shift of the cyclone track by 3-5° latitude to 60-65°N and high pressure conditions over Europe, similar to the warmest and driest summers of the period 1931-60. In winter, a similar pattern occured in the north, related to a blocking with severe winters and extended droughts, especially in eastern Europe.

After 1200 A.D., cooling in northern Greenland occured with a marked advance of glaciers in the Alps, a reappearance of Arctic sea ice, as well as extreme climatic anomalies and severe famines throughout Europe and a long drought period in Iowa and Illinois (Bryson and Murray, 1977): a transition towards the severe climate of the "Little Ice Age".

8. THE 20TH CENTURY WARMING TREND

The NH average temperature of this century reached a maximum around 1940, dropped by less than 0.5°C until about 1965 and has been slightly rising since. In fact, statistical analysis shows that the warming general trend of the 20th century started around 1900 and never reversed. The recent cooling of the NH which started about 1940 is only a manifestation of climatic fluctuations of higher frequency superimposed to this general trend (Goossens and Berger, 1983). The reference period 1931-60 was in fact one of the warmest periods over the last 500 years and was characterized by a relatively low interannual variability in some regions, but not in western and central Europe where severe winters and exceptional wet and dry summers were observed in the 1940's.

This general pattern differs however from place to place in the world. For example, a worldwide comparison for the period 1958-1976 has verified the warming trend also in higher southern latitudes but has indicated a cooling in northern southern temperate zones, as well as an increasing temporal variability in the tropics. This regionality in the climatic behaviour in time and space is confirmed by a recent statistical analysis of European stations (Goossens and Berger, 1983). It is remarkable that any tendancy can hardly be found in the Mediterranean stations (except for Lisbon and Palma).

Since all climatic variations are subject to large longitudinal variations, which can frequently mask the latitudinal changes, it is important to use maps for an adequate description of climatic change (Jones and Kelly, 1981-1982). Plausible patterns for temperature and precipitation changes accompanying a general global warming (such as might occur due to a large increase in atmospheric CO_2) have been analysed by comparing the five warmest years in the period 1925-74 with the five coldest in this period (Wigley et al., 1980). Temperature increases over most regions, except namely in south-eastern and south-western Europe. Precipitation changes are fairly evenly distributed between increases and decreases. Decrease occurs in western and central Europe, but an increase is observed in SE Mediterranean region. The general increase of temperature and decrease of precipitation could have considerable agricultural impact, but it may be pointed out that the reverse was observed for most of Greece and Turkey.

However, since the signal to noise ratio for the climatic response to CO_2 increase is highest in midlatitude summer or northern annual temperatures (Wigley and Jones, 1981) and since individual years can not take into account the subsequent changes in the oceans and cryosphere boundary conditions, the warmest (1934-53) and coolest (1901-20) twenty-year periods of this century based on Northern Hemisphere annual mean surface air temperature data have been compared as scenarios for Europe in a warmer world (Lough et al., 1983; Palutikoff et al., 1984). These periods differ by 0.4°C but show marked subregional scale differences from season to season. As compared to the coolest period, the warmest is characterized in Europe by a warming for the annual mean, spring, summer and autumn but winters over a large part of Europe are actually cooler : a belt of negative values extent from about 40° to 60°N and stretches across the whole of central Europe except to the extreme west, southern Spain and eastern Mediterranean. Winter temperature variability increases in Spain and southern France but decreases over the whole rest of the Mediterranean Basin. In summer, the whole central Mediterranean basin (Corsica to Crete) experiences an increase greater than 1°C but eastern Spain shows a cooling up to 0.5°C. These patterns are in fact related to higher mean sea level pressures in a warm world for most of Europe, except for the Mediterranean basin.

The effects of precipitation changes may be even more important for land ressources. In general, the warm period shows lower annual rainfall over most of Europe with a maximum decline in northern Italy and South central France where an increase in variability is also observed. Along the Mediterranean fringes, all the seasons are drier especially summer but Spain, Northern Italy and Yugoslavia receive more rain in winter, a season with a highly significant decrease in variability over the whole Mediterranean basin. An increase in variability associated with a decrease of rainfall as observed in Italy and probably in all the central part of the Mediterranean Basin both in summer and autumn of a warm Earth, would lead to a greater probability of drought.

9. SENSITIVITY EXPERIMENTS OF THE CLIMATE

9.1. Solar constant

Since the sun is the primary source of energy for driving the global atmospheric circulation, it is natural to consider possible variations in either the Sun's output or in the irradiation reaching the Earth, as likely to exert some control on the climate.

A simplified dynamical model of the global circulation (Wetherald and Manabe, 1975) indicates that a 2% increase in the solar constant would produce a rise of 3°C in the mean global surface temperature, but a decrease of 2% would produce an average temperature drop of 4.3°C. The induced changes are calculated to be much greater near the poles than at the equator because of the marked changes in snow cover and in albedo. The most marked effect was upon the precipitation where a 6% change (from -4% to 2%) in the solar constant produces a 27% increase in the area mean rate of precipitation. The magnetude of the change is particularly small in the subtropics and relatively large in middle latitudes of the model. The magnetude of the change in the evaporation rate varies less with respect to latitude than the change in precipitation rate with the exception of very high latitudes.

9.2. The Charney effect

9.2.1. Charney experiment. Since on average the Earth's land surface reflects back about 15% of the solar radiation, but with variations from 8% for dark green vegetation to about 80% for freshly fallen snow, widespread changes in vegetation, ice and snow cover or soil moisture could produce a significant change in the heat balance of the Earth and hence in climate.

Charney (1975,1977) has shown with the GCM of NASA Goddard Institute for Space Studies that increasing the albedo north of the intertropical convergence zone (ITCZ) from 14% to 35% has the effect of shifting this ITCZ several degrees of latitude south and so, decreasing the rainfall in the Sahel by about 40% during the rainy seasons. In fact, the vegetative cover at the border of a desert is to be regarded as being metastable, at least (i.e., an initial departure will tend to persist). The perturbation will thus remain until external counteracting climatological forces become large enough to overcome the local feedback effect and return the system towards the old equilibrium.

Charney et al. (1977) have shown that the physical mechanism which is responsible for the desert feeding back upon itself is the following : high albedo of a desert due to reduction of vegetation (at least with bio-geophysical feedback) → less absorption of solar radiation by the ground → less sensible and latent heat transfer to the atmosphere → less convective clouds → more solar radiation to reach the ground but even more reduction in downward flux of atmospheric long-wave radiation → net radiative heat loss relative to its surroundings further enhanced by the reduction in evapotranspiration (which is as important as changes in albedo) (desert is a radiative sink of heat). Since the ground stores little heat, it is the air that looses heat radiatively : horizontal temperature gradient appears and a frictionaly controlled thermal circulation which imports heat aloft is initiated with sinking motion and adiabatic compression to maintain thermal equilibrium.

As in the subtropics this sinking motion is superimposed on the descending branch of the mean Hadley circulation, but is even more intense, this seems to be a very important mechanism in semi-arid zones bordering on the major deserts.

As compared to the observed June-August mean rainfall, a model giving excessive evaporation over land and an increase of the desert and desert margins albedo from 0.14 to 0.35 leads to an increase of the rainfall over southern Europe by 30%, but there does not seem to be a difference between the low and high albedo cases. The evaporation is large over the Mediterranean and reaches 4 mm day for a precipitation ranging between 1 and 3 mm day^{-1}, indicating an important reduction in water availability.

9.2.2. <u>U.K. Met. Office experiment</u>. Using a simplified version of the U.K. Met. Office 11-layer model, Walker and Rowntree (1977) have also shown that ground dryness alone can cause deserts to persist. In the case of a dry Sahara with no soil moisture, crossing depressions produce little precipitation because there is no surface moisture to feed and maintain them. In a Sahara with a moist soil, the surface temperatures over the wet ground fall (relatively to the dry situation) by as much as 20°C and major depressions develop producing persistent widespread rain. Both dry and wet regions tend thus to be self-sustaining through the positive feedback effects of soil moisture, assisted by changes in vegetation cover and albedo.

Similar experiments are reported by Mintz (1981) where global models were run for northern summer conditions with land either all dry or all wet. Over much of the land, precipitation in the dry cases was found to be considerably reduced by absence of evaporation.

9.2.3. <u>Rowntree-Bolton experiment</u>. The experiments discussed by Rowntree and Bolton (1983) were designed to assess the response to an anomaly in the soil moisture of relatively small horizontal scale over Europe (similar to observed anomalies on the interannual time scale). The results show that such anomalies can have major effects on the modelled rainfall, humidity and temperature during the following 50 days over the anomaly area and that the anomalies can propagate into adjacent land areas. The northern regions of the Sahara and Italy have slightly more rainfall in the dry case but most North Africa and Southern Europe are drier. The differences in evaporation over the anomaly area are large (3 to 4 mm day^{-1}) and these changes extend well outside the anomaly area to the south over Mediterranean and mainly over Africa and Arabia.

The differences in temperature over the anomaly (higher in the dry case by 5 to 6°C over Europe) are clearly due to the local soil moisture, while the general warming over North Africa and Arabia in the dry case must be originally due to advection.

9.3. SST Anomalies

Persistent anomalies in ocean surface temperatures produce anomalies in the atmospheric circulation. Forcing of the atmosphere by the ocean is especially noticeable in the tropics but the influence of tropical sea surface temperature anomalies sometimes spread into middle latitudes. With the U.K. met. Office 5-layer model, Rowntree (1976) found that an anomaly of +2.5°C over a large area of the eastern tropical Atlantic Ocean produces an area of low surface pressure with a deficit of 7 mb centred west of the Bay of Biscay and an extensive area of high pressure with rises of up to 13 mb centered just east of Greenland. The modified circulation resulted in a strong easterly flow over the British Isles reminiscent of that which produced the very cold winter of 1962-63 in Britain.

9.4. Increasing Carbon Dioxide

The concentration of carbon dioxide in the atmosphere has increased by about 15% during this century and is currently rising at about 0.33% per year due largely to the burning of fossil fuels. Since it strongly absorbs the long wave radiation emitted by the Earth's surface, higher concentrations of carbon dioxide should produce higher temperatures in the troposphere by the so called "greenhouse effect" but, because the CO_2 in the stratosphere emits more infrared radiation to space than it absorbs, there should be a corresponding cooling of the stratosphere (Bach, 1982; Berger, 1984). This is confirmed by model calculations : for a doubling of the CO_2 concentration, Energy Balance Models simulate a surface warming of 1.3 to 3.3°C, Radiative Convective Models a warming of 1.3 to 3.2°C, and GCMs a warming of 2 to 3°C. Schlesinger (1983) has made a very extensive review of climate model simulations of CO_2 induced climatic change and Bach and Jung (this volume) have reviewed the simulated impacts on the european climates of such a CO_2 increase.

The GCMs simulations show that the CO_2-induced warming generally increases from the tropics to the poles, and also varies with longitude and with season. The increase in the average global surface temperature is 3°C, with a maximum of 10°C in polar regions caused partly by the retreat of the highly reflecting ice and snow surfaces and partly by the thermal stability of the lower troposphere limiting convective heat transfer to the lowest layers. In the tropics, this warming is spread throughout the entire troposphere by intense moist convection and so the temperature rise is smaller. Increased CO_2 also induces an increase in the global mean precipitation rate. The precipitation rate does not increase everywhere, however, and large decreases as well as increases are found in the tropics. A similar complex geographical pattern is found for the CO_2-induced changes in soil moisture. In many geographical locations there is a negative correlation between the changes in surface temperature and soil moisture, and a positive correlation between the changes in soil moisture and precipitation rate.

In the Northern Hemisphere, regions of large warming ($\Delta T > 4°C$) simulated by OSU model (Schlesinger, 1983) are located over Greenland, Arctic Ocean, northeast of Caspian Sea and the Sahara, Arabian and Gobi deserts. A region of large cooling is located in central East Africa where a large increase in precipitation rate and in soil moisture occurs. This model shows a remarkable increase by 2-3°C over western and southern Europe, which is found by the NCAR model only for $4xCO_2$.

It is also worth mentionning that the annually-averaged response of the troposphere in the GFDL simulation (Manabe, and Wetherald, 1980) with the annual insolation cycle is smaller than the tropospheric response with an annually-averaged insolation (but the stratospheric cooling is considerably larger). From a map of the annually averaged soil moisture distribution, it can be seen that a band of low soil moisture extends across the idealized continent between about 35° and 50° latitude, the maximum decrease being about 5 centimeters of water. This area of decrease of soil moisture can be attributed to the fact that the model's evaporation increased more than its precipitation for the warmer situation with doubled carbon dioxide.

When the climate model experiment with a mixed layer ocean was run to determine the change of conditions with season for a carbon dioxide doubling (Wetherald and Manabe, 1981), it was found that the maximum decrease of soil moisture at mid-latitudes occurred during the winter, spring and early summer. The band of dryness in this model experiment shifted poleward during this period. Another area of significant dryness in this model experiment appeared poleward of 60° during the early summer months. It should be noted that the dry period at both middle and high latitudes occurred during the season when adequate soil moisture is most needed by plants. Over most of Europe (western + central), in the $4xCO_2$ experiment, temperature increases by 5° to 8°C in winter but by less than 5°C in summer. In southern Europe, the seasonal difference is largely attenuated and temperature increases by 4°C all the year around.

In the $10xCO_2$ UK experiment (Mitchell, 1983), temperature increase is simulated everywhere in summer but is minimum over the eastern Mediterranean region where a decrease is even found in winter at a time western, central and northern Europe warms considerably (3 to 6°C).

The changes in zonal mean precipitation are positive almost everywhere in the GFDL and NCAR models but negative in the OSU model. A reliable geographical distributions of the change in precipitation rate is even much more difficult to obtain. In the seasonal GFDL model, annual mean precipitation will increase all over Europe for $4xCO_2$ except in the south-eastern regions. This is also found in the OSU model for $2xCO_2$, although the decrease extend much farther north into Central Europe in this case. A similar decrease in precipitation is also found in the $10xCO_2$ U.K. experiment both in winter and in summer, except for northern Europe.

An interesting experiment was also done by Mitchell (1983) where CO_2 is doubled and sea surface temperature is increased by 2°C (in order to take account the oceanic response to a CO_2 atmospheric warming) : the land surface temperature rises by 3°C, evaporation increases markedly over the oceans, precipitation increases in the main regions of atmospheric convergence and decreases in some regions of the subtropics, the mid-latitude peaks in precipitation have shifted poleward and the increases in the tropics are on the summer side of the equatorial peak. Temperature increases by roughly 2°C all the year around in southern Europe, whereas for central and northern Europe, temperature increases by 2°C in summer but by 5 to 10°C in winter. Pressure increases in western and southern Europe in summer (except in the eastern Mediterranean) and in winter. This is reflected in the seasonal changes of precipitation which decreases in summer all over Europe (except northern part) and in winter south of 48°N. In the NH, the largest area of reduced summer precipitation streches from Spain and North Africa across much of the Middle East and southern Asia into Mongolia, regions which are mostly arid in the control integrations. In general, the continents receive more precipitations where it is already heavy whereas arid regions receive even less precipitation. In winter, precipitation is more intense in the model's mid-latitude depression belt near 60°N. Over the continents, the drier areas are again confined mainly to regions which are arid in the control run; the main exception however is over southern Europe, where there is also a decrease in precipitation associated with a significant increase in the intensity of the model's east Atlantic anticyclone.

Further analysis indicates that, in general there is a negative correlation over land between the changes in surface air temperature and soil moisture; i.e. regions of maximum warming often occur where the soil moisture is decreased, particularly over the deserts, while regions of reduced warming or even cooling occur when the soil is moistened. There seems to be also a tendency for the change in soil moisture to be positively correlated with the change in precipitation rate, particularly over North and Central East Africa.

In the SST + 2xCO$_2$ Mitchell experiment, changes in P-E are positive in high latitudes and generally negative in low latitudes. Thus the warmer atmosphere picks up more moisture from the surface in low latitudes and returns more moisture to the surface in high latitudes. The zones of increase and decrease in P-E shift north and south following the seasonal migration of the sun. For example, over land in January between 0° and 40°N, there is a zone of reduced P-E; in August it is reduced to a narrow band at 45°N. In middle northern latitudes, the surface is drier most of the summer. Both the reduced summer precipitation and the removal of winter snow cover due to the general rise in temperature, contribute to the drier zone near 45°N in summer. Many of these changes are similar to those found by Manabe et al (1981).

In GFDL 2xCO$_2$ simulations, the soil moisture decreased almost everywhere poleward of 35° latitude and increased over most of the continent equatorwards of this latitudes. The maximum drying of the soil occurs in a band that stretches from coast to coast centered near 35 and 40° latitudes. This maximum, also found for the 4xCO$_2$ experiment a little bit north, is largely reduced when using a seasonal model.

Finally, the OSU model shows a small moistening of the soil over most of western Europe. A large drying of the soil is simulated for south-eastern Europe and North Africa.

10. CONCLUSIONS

With most of such examples of climatic variations, except for the well documented 20th Century, there is simply not enough regional and seasonal details known in order to use them for scenario development over a region the size of Mediterranean Basin or even Europe. Moreover, boundary conditions may have been substantially different from those of today. As, in addition, the models are strictly equilibrium response models and it is likely that the regional details of equilibrium and transient response models will differ at the scale of Mediterranean Basin, it is very difficult to draw detailed quantitative conclusions.

However, all the climatic situations presented here, clearly show that the climate of Europe and Mediterranean Basin and its sensitivity to the changes in the boundary conditions are highly variable. The vulnerability of Mediterranean countries to desertification will thus probably change as the climate changes progressively.

Keeping in mind the limitations of the paleoclimatic reconstructions and of the models themselves, we may tentatively conclude that the climate of the Mediterranean countries in southern Europe was more agreable (larger availability of water and/or milder summers) during the last Interglacial and the Climatic Optimum. For the 20th Century warm period and perhaps for the quite similar Medieval warm period (900 - 1300 A.D.), the western and eastern parts of Mediterranean Basin do not show similar patterns of precipitation and temperature anomalies, even for the annual average. Following the 5-year scenario by Wigley et al. (1980), only the eastern Mediterranean countries would benefit of a milder climate in a warmer Earth (the same conclusion can be drawn also from the Rowntree experiment on the european soil moisture anomaly).

On the other extreme side, the climatic situation in Mediterranean Basin, tends to have been worse during the Late Tertiary unipolar climatic asymmetry, the abortive glaciation 115,000 YBP, the Medieval warming (if we consider the higher frequency of blocking situations) and for a change of the albedo in the semi-desertic regions. For 18,000 YBP, east Mediterranean seems to have cooled by 1 to 2°C and have experienced a more positive balance between precipitation and evaporation.

REFERENCES

1. Adem, J., Berger, A., Gaspar, Ph., Pestiaux, P. and van Ypersele, J.P., 1984. Preliminary results on the simulation of climate during the last deglaciation with a thermodynamic model. In : "Milankovitch and Climate", Berger A., Imbrie J., Hays J., Kukla G., Saltzman B. (Eds), D. Reidel Publ. Company, Dordrecht (Holland), pp. 527-538.
2. Bach, W., 1982. Our Threatened Climate. D. Reidel Publ. Company, Dordrecht, Holland.
3. Bach, W. and Jung, H.J., 1986. The effects of model-generated climatic changes due to a CO_2 doubling on desertification processes in the Mediterranean area. This volume.
4. Berger, A.L., 1978. Long-term variations of daily insolation and Quaternary climatic changes. J. Atmos. Sci. 35(12), pp. 2362-2367.
5. Berger, A.L., 1979a. Spectrum of climatic variations and their causal mechanisms. Geophysical Surveys, 3, pp. 351-402.
6. Berger, A.L., 1979b. Insolation signatures of Quaternary climatic changes, Il Nuovo Cimento 2C(1), pp. 63-87.
7. Berger, A. (Ed.), 1981. Climatic Variations and Variability : Facts and Theories, Reidel Publishing Company, Dordrecht, Holland, 795pp.
8. Berger, A.L., 1983. Approach astronomique des variations paleoclimatiques : les variations mensuelles et en latitude de l'insolation de -130 000 à -100 000 et de -30 000 à aujourd'hui. Bull. Inst. Geol. Bassin d'Aquitaine, 34, pp. 7-26.
9. Berger, A., 1984. Man's impact on climate. In : "The Climate of Europe : Past, Present and Future, Natural and Man Induced Climatic Changes : an European Perspective", H. Flohn and R. Fantechi (Eds), pp. 134-197, D. Reidel Publ. Company, Dordrecht, Holland.
10. Berger, A.L., Imbrie, J., Hays, J., Kukla, G., saltzman, B. (Eds), 1984. Milankovitch and Climate. Reidel Publishing Company, Dordrecht (Holland).

11. Bryson, R.A. and Murray, Th.J., 1977. Climates of Hunger, University of Wisconsin Press, Madison.
12. Charney, J.G., 1975. Dynamics of deserts and drought in the Sahel. Quaterly Journal of the Royal Meteorological Society 101(428), pp. 193-202.
13. Charney, J., Quirk, W.J., Chow, S.H. and Kornfield, J., 1977. A comparative study of the effects of albedo change on drought in the semi-arid regions. J. Atmos. Sci. 34, pp. 1366-1385.
14. CLIMAP, 1976. The surface of the ice-age Earth. Science 191, pp. 1131-1137.
15. CLIMAP, 1981. Seasonal reconstructions of the Earth's surface at the last glacial maximum. Geological Society of America, Map and Chart Series MC-36.
16. Duplessy, J.Cl., Ruddiman, W.F., 1984. La fonte des calottes glaciaires. La Recherche 15, pp. 806-818.
17. Flohn, H., 1980. Possible Climatic Consequences of a man-made Global Warming. IIASA RR-80-30.
18. Frenzel, B., 1973. Climatic Fluctuations of the ice Age. Press of Case Western Reserve University, Cleveland and London.
19. Gates, W.L., 1976. Modeling the ice-age climate. Science 191, pp. 1138-1144.
20. Goossens, Chr., 1983. Evaluation of the statistical significance of climatic changes. In : "Second International Meeting on Statistical Climatology", pp. 14.6.1-14.6.8, Instituto Nacional de Meteorologia e Geofisica, Lisbon 26/09 - 30/09/83.
21. Goossens, Chr., 1985. Analysis of Mediterranean rainfall. J. of Climatology 5, in press.
22. Hammer, C.V., Clausen, H.B., Dansgaard, W., Gundestrup, N., Johnsen, S.J., Reeh, N., 1978. Dating of Greenland ice cores by flow models, isotopes, volcanic debris and continental dust. J. of Glaciology 20(82), pp. 3-26.
23. Hare, K., 1983. Climate and Desertification. World Climate program n°44, Geneva.
24. Jones, P.D. and Kelly, P.M., 1981-1982. Decadal surface temperature maps for the twentieth century. Parts 1 to 5. Climate Monitor 10(5), 11 (1 to 4). Climatic Research Unit, University of East Anglia, Norwich (England).
25. Kutzbach, J. and Guetter, P.T., 1984. Sensitivity of late glacial and Holocene climates to the combined effects of orbital parameter changes and lower boundary condition changes. Annals of Glaciology 5, in press.
26. Lamb, H., 1977. Climate : Present, Past and Future. Vol. 2, Methuen, London.
27. Landsberg, H.E. (Ed.), 1970. World Survey of Climatology, vol. 5-6, Elsevier Publ. Company.
28. Lorenz, E.N., 1967. The Nature and Theory of the General Circulation of the Atmosphere. World Meteorological Organization, Geneva.
29. Lough, J.M., Wigley, T.M.L., Palutikoff, J.P., 1983. Climate and climate impact scenarios for Europe in a warmer world. J. of Climate and Applied Meteorology, 22(10), pp. 1673-1684.
30. Manabe, S. and Hahn, D.G., 1977. Simulation of the Tropical Climate of an Ice Age. J. Geophysical Research 82(27), pp. 3889-3911.

31. Manabe, S., Wetherald, R.J., Stouffer, R.J., 1981. Summer dryness due to an increase of atmopsheric CO_2 concentration. Climatic Change 3(4), pp. 347-386.
32. Mason B.J., 1978. Recent advances in the numerical prediction of weather and climate. Proc. R. Soc. Lond. A. 36(3), pp. 297-333.
33. Mintz, Y., 1981. The sensitivity of numerically simulated climates to land surface conditions. JSC Study Conference on "Land Surface processes in Atmospheric General Circulation Models", WCRP, Geneva, pp. 109-114.
34. Mitchell, J.F.B., 1976-1977. Effect on climate of changing the earth's orbital parameters. Met. O. 20 Tech. Note n°II (72). Met. O. 20 Tech. Note n°II (100).
35. Mitchell, J.F.B., 1983. The seasonal response of a general circulation model to changes in CO_2 and sea temperatures. Quart. J. R. Met. Soc. 109, pp. 113-152.
36. National Research Council, 1982. Climate in Earth History. National Academy Press, Washington D.C.
37. Newson, R.L., 1973. Nature (London), 241, p. 39.
38. Nicholson, S.E. and Flohn, H., 1980. African environment and climatic changes, and the general atmospheric circulation in late Pleistocene and Holocene. Climatic Change 2(4), pp. 313-348.
39. Palutikoff, J.P., Wigley, T.M.L., Farmer, G., 1984. The impact of CO_2-induced climate change on crop yields in England and Wales. Progress in Biometeorology 3, pp. 320-334.
40. Petit-Maire, N., 1984. Le Sahara de la steppe au désert. La Recherche, 160, pp. 1372-1382.
41. Rowntree, P.R., 1976. Response of the atmosphere to a tropical Atlantic ocean temperature anomaly. Quaterly Journal of the Royal Meteorological Society 102, pp. 607-625.
42. Rowntree, P.R., Bolton, J.A., 1983. Simulation of the atmospheric response to soil moisture anomalies over Europe. Quaterly Journal of the Royal Meteorological Society 109, pp. 501-526.
43. Royer, J.F., Deque, M., Pestiaux, P., 1983. Orbital forcing of the inception of the Laurentide ice-sheet ? A GCM simulation of the 125,000 -115,000 BP transition. Nature 304, pp. 43-45.
44. Royer, J.F., Pestiaux, P., 1984. A sensitivity experiment to astronomical forcing with a spectral GCM for the simulation of July 125 kyr BP. In : "New Perspectives in Climate Modelling", Berger A.L. and Nicolis C. (Eds), Elsevier Publ. Company pp. 269-286.
45. Saltzman, B. and Vernekar, A.D., 1975. A solution for the Northern Hemisphere Climatic Zonation during a Glacial Maximum. Quaternary Research 5, pp. 307-320.
46. Schlesinger, M.E., 1983. A review of climate model simulations of CO_2-induced climatic change. Climatic Research Institute Report n°41. Oregon State University.
47. Street, F.A. and Crove, A.T., 1976. Environmental and climatic implications of late Quaternary lake-level fluctuations in Africa. Nature 261, pp. 385-390.
48. Walker, J. and Rowntree, P.R., 1977. The effect of soil moisture on circulation and rainfall in a tropical model. Quaterly Journal of the Royal Meteorological Society 103(435), pp. 29-46.
49. Wetherald, R.J. and Manabe, S., 1975. The effects of changing the solar constant on the climate of a General Circulation Model. J. Atmos. Sci. 32, pp. 2044-2059.

50. Wigley, T.M.L., 1977. Geographical Patterns of Climatic Change 1000 BC
    - 1700 AD. NOAA Report 7-35207.
51. Wigley, T.M.L., Jones, P.D., Kelly, P.M., 1980. Scenario for a warm
    high CO_2 world. Nature 283, pp. 17-21.
52. Wigley, T.M.L., Jone, P.D., 1981. Detecting CO_2-induced climatic
    change. Nature 292, pp. 205-207.
53. Williams, J., Barry, R.G. and Washington, W.M., 1974. Simulation of
    the atmospheric circulation using the NCAR Global Circulation Model
    with Ice Age boundary conditions. J. of Applied Meteorology 13(3),
    pp. 305-317.
54. Woillard, G., 1979. Abrupt end of the last interglacial s.s. in
    north-east France. Nature 281, pp. 558-562.

THE EFFECTS OF MODEL-GENERATED CLIMATIC CHANGES DUE TO A CO_2 DOUBLING ON DESERTIFICATION PROCESSES IN THE MEDITERRANEAN AREA

H.J. JUNG and W. BACH
Center for Applied Climatology and Environmental Studies,
Dept. of Geography, University of Münster,
D-4400 Münster, West-Germany

Summary

For the estimation of a climatic change induced by a doubling of atmospheric CO_2 we use the results of three-dimensional general circulation models (GCM). Although the results from present climate modeling cannot be considered as predictions of future climatic conditions due to the inherent models' deficits, they can still serve a useful purpose in climate change scenarios. The reason for this is that climate models are the only tools available to study the response of the climate system to a perturbation in a physically consistent manner and that such types of models can provide a consistent data set of high temporal and spatial resolution. For the Mediterranean area, the results obtained from three different GCMs, namely, the British Meteorological Office model (BMO), the Goddard Institute of Space Studies model (GISS), and the National Center for Atmospheric Research model (NCAR) are shown. The regional and seasonal distributions of temperature, precipitation, and soil moisture are used to study the potential for desertification. The results indicate that the CO_2-induced changes for temperature generated by the three models are quite similar. The values of the area mean change range between 2.5 and 4.2 K. The precipitation response results in a diverse pattern. The physical mechanisms likely to be responsible for the climatic changes are identified and their statistical significance is tested. This type of work will help us develop the methodology and assist us in gaining insight into the use of climate model scenarios for impact analysis.

1. INTRODUCTION

Desertification in the Mediterranean area depends strongly on climate because this region belongs to the semi-arid climatic zone. Changes in the mean values and the interannual and seasonal variability of climate parameters, such as temperature and precipitation, exert a strong effect on water balance, soil conditions, and vegetation etc. These processes, which may induce desertification, are often amplified by man's activities such as through intensive agriculture, livestock and deforestation. Man is now in a position not only to change the environmental conditions but he can also influence the climate directly by increasing the atmospheric CO_2 concentration through the massive use of fossil fuels.

Three-dimensional general circulation models (GCMs) based on physical laws governing the atmosphere's structure and behavior have been used to study the climatic changes likely to result from increased levels of CO_2.

A recent re-evaluation of GCM simulations published by the National Academy of Sciences (18) gives a global temperature increase at the surface of 3 ± 1.5 K for a doubling of CO_2. Experiments with the most sophisticated GCMs indicate that the response of climate to a CO_2 increase will not be uniform in space and time, but will rather show pronounced seasonal and regional inhomogeneities. Thus, the analysis of a climatic impact on desertification, agriculture and energy etc. requires a description of the regional and seasonal distribution of a simulated climatic change, a description of the degree of uncertainty of the simulated climate parameter, and a transformation of the climatic change from the model's large scale onto the smaller local scale on which climatic impact occurs (2, 5). At present no GCM is capable of providing a definitive prediction of regional climatic change. Therefore, climatic change scenarios have to be developed as useful tools for impact analysis. These can also be constructed from historical or paleoclimatic data. For these "analog studies", periods of the earth's climate history can be examined when the globally-averaged surface air temperature was higher than today (10). These climatic states are, however, not necessarily related to atmospheric CO_2 concentration. Therefore, climate models must be used because they are the only available tool at present to study, in a physically consistent manner, the climatic change resulting from a future CO_2 increase. The results from different climate models are compared and then used to develop climatic change scenarios.

2. DEVELOPMENT OF CLIMATIC CHANGE SCENARIOS

The climatic change scenarios developed for this report are intended to provide seasonal and regional distributions of temperature, precipitation, and soil moisture for the study area (i.e. 10° W - 35° E and 30° N - 50° N). These are derived with the purpose of showing the potential structure and direction of a climatic change resulting from a CO_2 doubling. The climatic change scenarios should neither be construed as predictions of a future climatic change caused by changes in CO_2 concentrations and other boundary conditions, nor as estimates of forthcoming climatic events due to the inherent variability of climate, but rather as a set of self-consistent and plausible patterns of change (1).

2.1 General circulation models

In order to study the effects of a CO_2 doubling, we make use of sensitivity experiments that have been performed with GCMs at the British Meteorological Office, Bracknell, UK (BMO); the Goddard Institute for Space Studies, New York, USA (GISS) and the National Center for Atmospheric Research, Boulder, USA (NCAR). For an overview on present CO_2 simulations see e.g. Schlesinger (17).

As noted earlier, GCMs are based on the fundamental dynamical equations which describe large-scale atmospheric circulation. The prognostic variables wind velocity, temperature, pressure, density, and humidity are determined by the equations of motion, the first law of thermodynamics, the continuity equations for mass and water vapor, and the equation of state. In addition, ground temperature, soil moisture, and mass of snow on the ground are computed by energy, water and snow budget equations. Furthermore, quantities such as the transfer of solar and terrestrial radiation, the condensation of water vapor, the formation of clouds and their radiative interaction, as well as the turbulent transfer of heat, moisture and momentum are determined from parametrizations which relate these sub-grid scale processes to the large-scale variables resolved by the model. Moreover, boundary conditions at the earth's surface including the sea surface temperature, the sea ice distribution, the land surface elevation and the surface albedo distribution are specified. The solar radiation at the top of the atmosphere is either held fixed or it is assigned the normal daily or seasonal variation.

Characteristic properties of the GCMs presented in this paper are shown in Table I. The models differ with respect to their spatial resolution, their ocean/sea ice model, their cloud model as well as their treatment of soil moisture. The horizontal domain of all models is global, that is, 360° of longitude and extending from pole to pole for a realistic land-ocean distribution and topography. In the vertical the models have 5 to 9 layers including both the troposphere and the stratosphere.

In modeling the ocean and sea ice effects on climate, two different concepts are used. In the BMO-GCM, the sea surface temperatures (SST) and the sea ice distributions are prescribed to be equal to their observed values and to follow seasonal cycles. Thus, no feedback processes and interactions can take place between the atmosphere and the ocean. In the 2 x CO_2 simulation, the SST is enhanced at all ocean gridpoints by 2 K to account for the increase in SST due to a CO_2 doubling found in other CO_2 GCM experiments. In the second approach, namely the GISS-GCM, a mixed layer concept is used to compute ocean temperatures and ice cover based on energy exchange with the atmosphere, ocean heat transport, and the ocean mixed layer heat capacity. The latter two are specified, but vary seasonally at each gridpoint. The NCAR-model, on the other hand, has fixed mixed layer depths and no horizontal heat transport. CO_2 - sensitivity experiments with fully coupled atmosphere - ocean general circulation models are in preparation.

A cloud prediction scheme is incorporated in the simulations of NCAR and GISS. Prescribed distributions of seasonally-varying cloudiness are used in the BMO experiments.

The treatment of soil moisture in the BMO and NCAR-GCM is based on Manabe (11). The soil moisture content is a function of rainfall, condensation, evaporation and snowmelt. The maximum soil moisture content amounts to 20 cm. In the GISS-GCM a two-layer model is used with a variable water storage capacity depending on vegetation type.

The governing equations of the GCMs are solved by numerical methods. The variables at the grid points are integrated forward in time until the model approaches a statistical equilibrium. The time required varies from ≈ 3 years (BMO) to 30 years (GISS) depending on the type of ocean model.

2.2 Analysis of GCM results

The output from GCMs for impact analysis considered here includes the temporal and spatial distribution of temperature, precipitation and soil moisture. A comparison between the results of different GCMs can only be performed effectively, if the data of all models are transferred to a reference grid system. This grid (4° lat. x 5° long.) is used for all data processing, i.e. spatial and temporal filtering to remove small-scale fluctuations. For the spatial filtering a Gaussian filter is used, whereas for the temporal filtering a fourth order binominal filter is applied.

For the interpretation of the CO_2 sensitivity experiments it is necessary to determine whether the changes obtained are significant in a statistical sense. The difference between the 2 x CO_2 and the 1 x CO_2 experiment, the 'signal', arises mainly from changes in the boundary conditions. This signal may be obscured by 'noise' inherent in the model. If the signal-to-noise ratio is statistically significant, then the new equilibrium climate for 2 x CO_2 can be considered as different from that of the 1 x CO_2 experiment. Therefore, the statistical significance of the

Table I Characteristics of general circulation models used in the simulation of a CO_2-induced climatic change

Model characteristics	GISS	BMO	NCAR
Temporal resolution	annual cycle	annual cycle	seasonal
Spatial resolution	V: 7 layers H: $\Delta\Phi$ = 8° $\Delta\lambda$ = 10°	V: 5 layers H: $\Delta\Phi$ = 3° $\Delta\lambda \approx$ 330 km	V: 9 layers H: $\Delta\Phi$ = 4° $\Delta\lambda$ = 7.5°
Land-ocean distribution	realistic	realistic	realistic
Topography	realistic	realistic	realistic
Ocean/sea ice	mixed layer with prescribed seasonal depth and horizontal heat transport/ice thickness predicted	ocean surface temperature and ice cover prescribed/surface temperature + 2 K	50 m mixed layer (no ocean heat transport)/ice cover predicted
Land temperature	predicted	predicted	predicted
Soil moisture	predicted	predicted	predicted
Clouds	predicted when cumulus convection or relat. humidity $\geq$ 100%	prescribed annual cycle	predicted
Global temperature change for a CO_2 doubling	4.0 K	2.2 K	4.2 K
Global precipitation change for a CO_2 doubling	0.4 mm/day	0.14 mm/day	

V: vertical resolution; H: horizontal resolution, Φ, λ: latitude, longitude

BMO : British Meteorological Office, Bracknell, U.K. (13)
GISS: Goddard Institute for Space Studies, New York, U.S.A. (7,8)
NCAR: National Center for Atmospheric Research, Boulder, U.S.A. (19)

changes between the two model runs must be determined before we can derive
meaningful climate scenarios. Since the time series simulated by GCMs are
highly correlated in space and time the standard methodology of statistical
inference can not be applied. The principal difficulty is to obtain an
estimate of the variance of time averages. Katz (9) uses a parametric time
series model to estimate the variance of daily averages, whereas the method
of Chervin (3), which we applied here, is based on a conventional variance
estimate using monthly means for consecutive years. The null hypothesis
that there is no difference between the control and the perturbed experiment is tested at each gridpoint by computing the test variate:

$$r = \frac{<m_C> - <m_E>}{\left(\sigma_C^2 + \sigma_E^2\right)^{1/2}}$$

where $<m_C>$ and $<m_E>$ are the estimated ensemble means of the control
$(1 \times CO_2)$ and the experiment $(2 \times CO_2)$. The acceptance region for the
null hypothesis is

$$r_1(min) < r < r_2(max)$$

where $r_1(min)$ and $r_2(max)$ are determined from modified values of Student's
t-distribution.

A comparison of the regional climatic change distributions obtained
from different models can help us identify the regional effects of a CO_2-
induced change. The agreement between different patterns are calculated
for the deviations from the area average at every gridpoint. From these
deviations the number of points with agreement of sign, and the correlation
coefficient of the deviations are computed. The first measure gives the
overlapping areas of the two data sets with values below or above the respective area mean, whereas the second describes the amplitudes of the
deviations.

2.3 Simulation of present climate

For the use of climate models in regional impact analysis the model
climate must be verified on a regional scale by comparing the computed
with the observed climate parameters. On this scale differences between
the two data sets can be caused not only by the imperfect performance
of the model, but also by the use of different grid systems of the two data
sets, and by factors related to data collection. For model validation we
will compare the regional distribution of annual mean temperature and
annual mean precipitation rate for the GISS control experiments with the
respective measured data obtained by W.L. Gates. Figure 1a shows the temperature difference between the GISS control experiment and the observed
data. In the southern as well as the northern parts of the study area the
model-generated temperature is generally lower than the observed data
reaching a maximum value of -2 K. There are also large areas in the center
of the study area with deviations near zero.

The annual mean precipitation rate of the GISS-experiment is considerably higher than that of the measured global average (17). This finding is
also valid for our study area. Here a large area with deviations greater
than 1 mm/day covers the Mediterranean and south-western Europe. It should
be noted that the magnitude of this change is comparable to the annual mean
precipitation at several stations in this area. These discrepancies, as
large as a factor of two, are probably due to the GCM modeling because the
precipitation is both the result of large-scale vertical motion and of con-

a   b

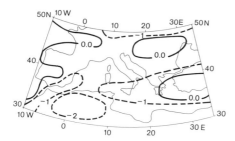 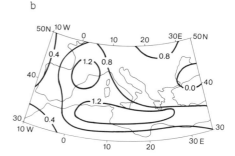

Figure 1a, b   Regional distribution of the differences between the GISS con-
               trol experiment and the measured climate for the annual mean.
               a: temperature (K); b: precipitation rate (mm/day)

vective motion on a local scale below the GCM resolution. Therefore, the
latter type of precipitation is parametrized in terms of the large-scale
distribution of temperature and humidity.
     A model verification scheme should also include a comparison between
the simulated and observed climatic variability on annual and seasonal
scales. This internal variability of GCMs results primarily from the simulation of the unstable transient cyclones in middle latitudes (5). We have
not done this type of model validation because we did not have the data on
the observed variability.

2.4 Regional and seasonal simulation of a CO_2- induced climatic change

2.4.1 Temperature response
     The regional distribution of temperature change (2 x CO_2 - 1 x CO_2)
for both the annual mean and the winter and summer seasons is shown for
the GISS-GCM in Fig. 2 a-c. The annual mean temperature change is 4 to 5 K.
Values above 4 K are found over North Africa. In winter there is a zonal
distribution with values exceeding 4 K in the northern part of the study
area. The distribution in summer shows a temperature increase of 2 - 3 K
in central and eastern Europe and one exceeding 4 K over the southern
Mediterranean and western parts of North Africa. Statistical tests were
only performed for the seasons showing significant changes at the 5% level
everywhere except for a small area covering Greece and Bulgaria in summer.
     For the BMO-GCM (Fig. 2 d-f) an annual mean temperature change of 4 K
is found over western North Africa. Over the remaining areas the increase
amounts to 3 K. The winter situation is characterized by a strong temperature increase over central and eastern Europe which is related to increased
westerly flow and the reduction in snow cover. In summer high temperature
increases occur over western North Africa and southern Spain as a result
of the poleward shift of the subtropical high pressure belt with a resulting
reduction of precipitation and soil moisture (13). In winter only the
western-most fringe of the study area shows significant changes. The changes
are significant everywhere in summer.
     The temperature distribution computed by the NCAR-GCM for the winter
season shows somewhat of a meridional structure with values above 4 K to
the west of Italy and the east of Greece (Fig. 3 a). In summer the tem-

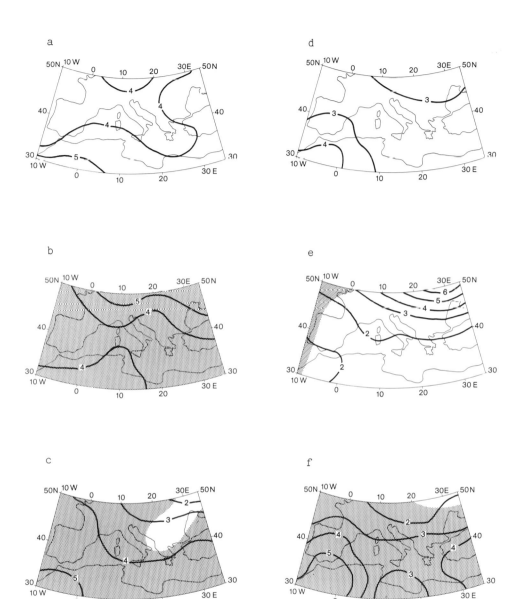

Fig. 2a - f Regional distribution of the temperature change (K) (2 x CO_2 -
1 x CO_2) for the GISS (a - c) and the BMO (d - f) experiment.
Stippling indicates statistical significance (5% level).
a,d: annual mean; b,e: winter; c,f: summer

perature increase is smaller than 4 K almost everywhere except for a small part over North Africa and Spain (Fig. 3b). In both seasons the warming due to increased CO_2 is significant at the 5% level.

a   b

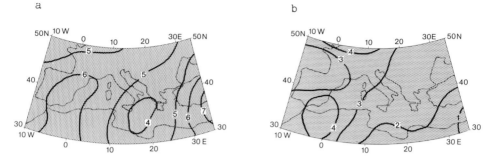

Fig. 3a, b   Regional distribution of the temperature change (K) (2 x CO_2 -
             1 x CO_2) for the NCAR experiments. Stippling indicates statis-
             tical significance (5% level).
             a: winter; b: summer

2.4.2 Hydrological response

The change in hydrological parameters such as precipitation and soil moisture is responsible for amplifying or dampening desertification processes. Fig. 4 a-c shows the differences of precipitation rates of the 2 x CO_2 minus control GISS simulation runs for both the annual and the winter and summer situations. The mean annual deviations are negative in the south-eastern part of the study area. An increase of 0.4 mm/day is found over northern Italy and the Alps. In winter, a precipitation rate decrease of 0.2 mm/day is found in two small areas in the western and central Mediterranean. The northern parts of the study area show a precipitation rate increase. In summer, an area of decrease is found over northern Spain and south-western France and in the eastern Mediterranean. High simulated changes exceed the observed values over North Africa. None of the changes is significant.

Regional distributions of the precipitation changes simulated by the BMO-GCM are shown in Fig. 4 d-f. For the annual case precipitation decreases by 0.4 - 0.8 mm/day south of 50° N. A similar situation is found for winter with maxima up to 1.2 mm/day over the Biscay and the Balkans. The summer pattern shows areas of high decreases (-1.2 mm/day) centered over North Africa and Yugoslavia. Significant changes at the 5% level are stippled.

The regional distribution of precipitation changes computed by the NCAR-GCM is shown in Fig. 5a and b. In winter, the area of decrease is located over much of Portugal and Spain. Positive changes are shown over the southern part of central Europe. In summer, positive differences of the precipitation rate prevail. The largest changes occur east of 15°. In contrast to temperature, the areas with significant precipitation changes are smaller.

The GISS, BMO and NCAR models have been used to simulate also the regional distribution of soil moisture changes for January and July. Since the parametrization schemes applied in these models are different, only the direction of changes can be indicated. The soil moisture changes reflect the precipitation changes. The GISS simulation shows an increase over central Europe and Libya in winter (Fig. 6a) and a similar pattern in summer (Fig.6b).

a
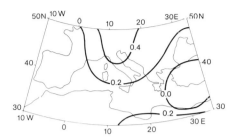

d
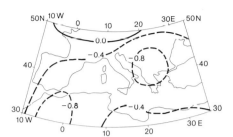

b
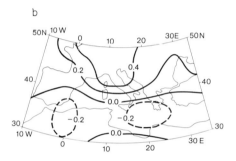

e
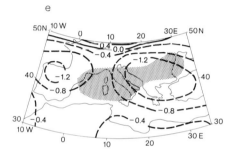

c
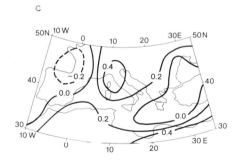

f
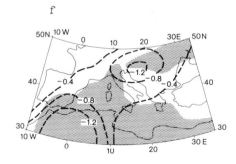

Fig. 4a - f Regional distribution of the precipitation rate change (mm/day)
(2 x CO_2 - 1 x CO_2) for the GISS (a - c) and the BMO (d - f)
experiments. Stippling indicates statistical significance
(5% level).
a,d: annual mean; b,e: winter; c,f: summer

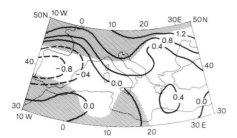
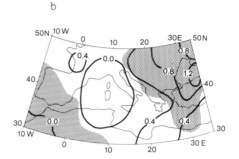

Fig. 5a, b Regional distribution of the precipitation rate change (mm/day)
(2 x CO_2 - 1 x CO_2) for the NCAR experiment. Stippling indicates
statistical significance (5% level).
a: winter; b: summer

As a result of the decrease in winter precipitation, the soil moisture
of southern central Europe is also reduced in the BMO simulation (Fig. 6c).
In summer, the surface is much drier. This increased summer dryness, which
is also found by Manabe et al. (12) in their integrations can be attributed
to both the reduced summer distribution of precipitation and the general
rise in temperature leading to an earlier removal of winter snow (13).

In the NCAR simulations, the greater precipitation in winter (Fig. 6e)
results in an increase in soil moisture over central Europe whereas North
Africa shows a small decrease. In summer, there is a decrease over eastern
France and patches over North Africa (Fig. 6f). It is worthwhile noting
that areas of heavy precipitation increases (eastern Europe) coincide with
those of high soil moisture. Soil moisture is more variable than temperature.
Thus, areas with significant changes in soil moisture are much smaller than
those for surface air temperature. Additional studies about the rate of
change in hydrological parameters, especially in soil moisture, have been
conducted by Rind (15), Mitchell (14), as well as Rowntree and Bolton (16).

2.4.3. Discussion

The CO_2-induced changes of temperature and precipitation shown in
the previous sections differ for the three GCMs because the models exhibit
differences in terms of spatial and temporal resolution, parametrization
of physical processes, and representation of the ocean. Since it is not
possible to determine, at present, which of the models is more accurate,
it is useful to compare the results of different models in terms of the
sign of the predicted change and the correlation coefficient. Table II
shows the respective values for the BMO and GISS temperature and precipitation distributions. The areal mean changes for temperature are quite similar.
This is in contrast to precipitation where the values are of opposite
sign. This feature is also visible in the correlation coefficient for July.

Since the globally-averaged warming (see Table I) disagrees by a
factor of two, it is not surprising that also the regional changes are
quite uncertain. The CO_2-induced changes of local precipitation are even
more uncertain than those of temperature because there is a large disagreement among the GCMs even for the zonally-averaged changes. This is due to
the fact that precipitation shows a relatively large temporal and spatial
variability due to convective-scale processes. Since the identification of

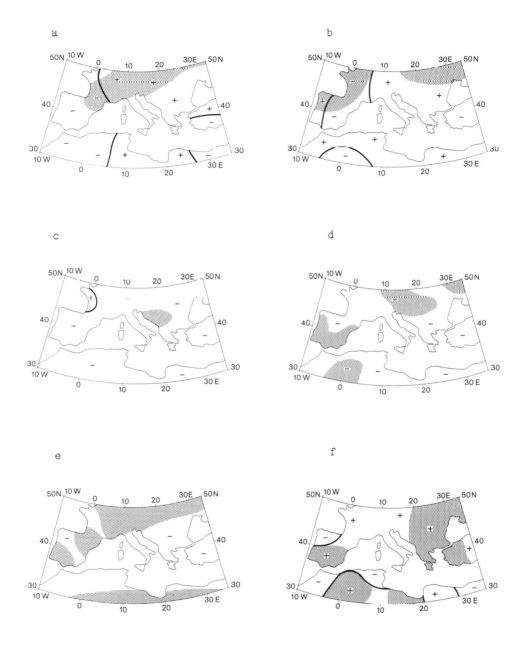

Fig. 6a - f Regional distribution of the soil moisture change (2 x CO_2 - 1 x CO_2) for the GISS (a, b), the BMO (c, d) and the NCAR (e, f) experiments. Stippling indicates statistical significance (5% level).
a, c, e: winter; b, d, f: summer

Table II   Comparison of the BMO and GISS experiments (2 x CO_2 - 1 x CO_2)

Temperature

Season	Experiments	Areal mean change (K)	Number of points (%) with agreement of sign	Correlation coefficient
Annual	GISS	4.1	66	0.5
	BMO	3.0		
January	GISS	4.2	86	0.8
	BMO	2.5		
July	GISS	3.9	75	0.7
	BMO	3.3		

Precipitation

Season	Experiments	Areal mean change (mm/day)	Number of points (%) with agreement of sign	Correlation coefficient
Annual	GISS	0.2	66	0.3
	BMO	-0.4		
January	GISS	0.1	35	0.2
	BMO	-0.6		
July	GISS	0.1	25	-0.3
	BMO	-0.5		

regional changes is a priority task for climate impact assessment, those
regions in which the simulated changes agree, have to be determined. Changes
in areas with agreement of sign can be considered as reliable, if
    - each model simulates quite well the observed climate,
    - the predicted climatic change for a CO_2 doubling can be explained
      in terms of physical processes, and if
    - the response of each model is statistically significant (2).
    Compared to the areal averages, the temperature distributions ob-
tained from the three models show similar results, namely,
    - higher values in north-eastern Europe in winter and in summer,
    and
    - higher values in south-western Europe and North Africa.
    The simulated precipitation changes are often not significant and
the only reliable statement that can be made at this stage is that the
hydrological cycle will be enhanced for the global average (18). Therefore, we will consider here only the longitudinally-averaged distribution,
which shows an above average increase north of 40° (GISS) and 48° (BMO)
in winter. The distributions for summer are completely different for these
two models. The causes for these differences on a local scale may be attributed to the different parametrizations of the many climatic processes
involved. To understand the causes for these differences, it would be
necessary to analyse for each model the treatment of the various sub-grid
scale processes which include convection, cloud formation and precipitation.
The disparity in the resolution of the models may be another reason for the
disagreement in generated climates.
    At present, GCM results can be considered to give a plausible pattern
of climatic change due to a CO_2 doubling and other changes in boundary con-

ditions. Therefore, it would be appropriate to assess, in a next step, the effects of climatic changes on ecosystems, using e.g. crop-yield models or biomass productivity models (4). Results obtained from these studies could give both an indication of the sensitivity of the ecosystem to climatic change and an idea which model-generated climate parameters would be most useful in impact studies.

3. CONCLUSIONS

Three-dimensional climate models (GCMs) have been used to study the regional and seasonal response of temperature, precipitation and soil moisture distributions to a CO_2 doubling. The results of different GCMs are applied in order to increase the reliability of climatic change scenarios, which are useful tools for climatic impact assessment in such areas as desertification, agriculture, water resources, and energy use. GCMs are not only capable of simulating the climatic response due to a perturbation in a physically consistent manner, but they can also provide a comprehensive set of variables which is internally consistent. However, it must be realized that the ultimate value of a model depends upon its accuracy in comparison with observations. Therefore, it is of limited value to simulate a climate change with a high statistical significance, if the model's climate itself has unacceptably large errors (5). GCMs also simulate the climate variability on a global scale. Statistics in terms of frequency distributions and scatter diagrams may provide additional information about climatic change (6). Parameters such as the number of days without precipitation or the frequency of extreme events are particularly important for local ecosystem studies.

No doubt such studies as described above are very useful in that they can help improve the methodology for present GCMs to perform more adequately on a regional scale. Work is under way to transfer the present model's global-scale outputs to smaller scales (5). Such improvements will allow results from climate models to be used with greater confidence in impact models, such as crop-yield and biomass productivity models, for the purpose of studying the potential effects of climatic changes on the ecosystem.

ACKNOWLEDGEMENTS

We wish to thank Lawrence Gates, Oregon State University, for making available the observed data, and Warren Washington, National Center for Atmospheric Research, and James Hansen, Goddard Institute of Space Studies, for supplying the model-generated data. Special thanks go to John Mitchell, British Meteorological Office, for both supplying data and reviewing this paper.

REFERENCES

1. BACH, W. (1984). Our threatened climate: Ways of averting the CO_2 - problem through rational energy use, Reidel Publ. Co., Dordrecht
2. BACH, W., JUNG, H.J. and KNOTTENBERG, H. (1984). Developments of regional climate scenarios. In: Socioeconomic impacts of climatic changes, Report for the EEC and the BMFT, Dornier System, Friedrichshafen
3. CHERVIN, R.M. (1981). On the comparison of observed and GCM simulated climate ensembles, J. Atmos. Sci. 38, 885 - 901
4. DENNETT, M.O., ELSTON, J. and DIEGOQ, R. (1980). Weather and yields of tobacco, sugar beet and wheat in Europe, Agr. Meteor., 21, 249 - 263
5. GATES, W.L. (1983). The use of general circulation models in the analysis of the ecosystem impacts of a climatic change. Paper presented

at the Study Conference on the sensitivity of ecosystem and society to climatic change, Villach, 19-23 September, 1983

6. GATES, W.L. and BACH, W. (1981). Analysis of a model simulated climate change as a scenario for impact studies. Report for the German Federal Environmental Agency, R&D No. 104-02-513, 163pp
7. HANSEN, J., RUSSELL, G., RIND, D., STONE, P., LACIS, A., LEBEDEFF, S., RUEDY, R. and TRAVIS, L. (1983). Efficient three dimensional global models for climate studies: Models I and II, Mon. Wea. Rev. 110, 609 - 662
8. HANSEN, J., LACIS, A., RIND, D., RUSSELL, G., STONE, P., FUNG, J., LERNER, J., RUEDY, R. (1984). Climate sensitivity experiments with a three-dimensional model: Analysis of feedback mechanisms (unpublished manuscript)
9. KATZ, R.W. (1982). Statistical evaluation of climate experiments with general circulation models: A parametric time series modeling approach, J. Atmos. Sci. 39, 1446 - 1455
10. LOUGH, J.M., WIGLEY, T.M.L. and PALUTIKOF, J.P. (1983). Climate and climate impact scenarios for Europe in a warmer world, J. Cl. and Appl. Meteor., 22, 1673 - 1684
11. MANABE, S. (1969). Climate and the ocean circulation I: the atmospheric circulation and the hydrology of the earth's surface. Mon. Wea. Rev., 97, 739 - 774
12. MANABE, S., WETHERALD, R.T. and STOUFFER, R.J. (1981). Summer dryness due to an increase of atmospheric CO_2 concentration, Climatic Change 3, 347 - 386
13. MITCHELL, J.F.B. (1983a). The seasonal response of a general circulation model to changes in CO_2 and sea surface temperature, Q.J.R. Met. Soc. 109, 113 - 152
14. MITCHELL, J.F.B. (1983b). The hydrological cycle as simulated by an atmospheric general circulation model. In: A. Street-Perrott et al. (eds.) Variations in the global water budget, 429 - 446, Reidel Publ. Co., Dordrecht
15. RIND, D. (1982). The influence of ground moisture conditions in North America on summer climate as modeled in the GISS GCM, Mon. Wea. Rev., 110, 1487 - 1494
16. ROWNTREE, P.R. and BOLTON, J.A. (1983). Effects of soil moisture anomalies over Europe in summer. In: A. Street-Perrott et al. (eds.) Variations in the global water budget, 447 - 462, Reidel Publ. Co., Dordrecht
17. SCHLESINGER, M.E. (1983). Simulating CO_2-induced climatic change with mathematical climate models: Capabilities, limitations and prospects, III3 - III139, US DOE 021, Washington, D.C.
18. US NRC (1983). Changing climate, National Academy Press., Washington, D.C.
19. WASHINGTON, W.M. and MEEHL, G.A. (1984). Seasonal cycle experiments on the climate sensitivity due to a doubling of CO_2 with an atmospheric general circulation model coupled to a simple mixed layer ocean model, J. Geophys. Res. (in press).

LANDSCAPE CHANGES IN GREECE AS A RESULT OF CHANGING CLIMATE DURING THE QUATERNARY

R. PAEPE
Belgian Geological Survey
and
Vrije Universiteit Brussel

Summary

Eversince the Tertiary, landscape development underwent in Greece Typical Savanna-tropical conditions during the Early Pleistocene; semi-arid piedmont, and glacis development during the Middle Pleistocene and dry-desertic conditions during the later part of the Middle and the whole of the Late Pleistocene.
During the Holocene not less than 19 holocene soils occurred to exist interfering with fluviatile phases. As of 700 BC valleys are completely filled up, so that also present coastal configuration came into being.
Moreover after 700 BC, four dry phases occurred every 1000 years namely : 8th Cent. B.C., Middle to Late Roman, Middle Byzantine as well as todays desertification trend.

1. INTRODUCTION.

Although zonal distribution of climato-geographical belts along the Earth's surface from Equator to Poles is a long time established concept it still remains difficult to imagine that such zonation changed as much rapidly as climatic fluctuations started to occur eversince Pliocene times. The consequences of such climatic changes and related shifts of the climatic belts along the Earth's surface are multiple.

Indeed similar and simultaneously to the wandering of the vegetation cover from Equator to Poles during the warm periods and vice-versa during the cold periods, landscapes moved along the globe's parallels. Hence and forth relicts were left and inherited by the next to come landscape.
As a direct result, one is witnessing in a given point at a given moment of a mixture of inherited landscape features which are usually quite difficult to disentangle through space and time.
With regard to such evolution through time special attention must be paid to the periodicities characterising such changes. First of all from the geodata it occurs that long term and short term cycles interfere in a quite complicated manner. Moreover, intensities of such varations differ and often show unexpected relations either to long or to short term cycles. This is a further complication which makes the reading of old landscapes still more difficult. For example, tropical landscapes which extended up to 70°N. till the end of the Miocene shrinked their extension to latitudes below 23°N. during the Quaternary cooling. Even today within a warm period similar to an interglacial their northwards extending impact is fading out rather quickly.
From this concept, the Greek scenery will be examined in view of a search for long term and short term climatic variations which may have conditioned the landscape machineric leading to the "mixed subtropical landscape" of Greece and surrounding countries.

2. THE "LANDSCAPE MACHINERIE"

2.1. MIXED SUBTROPICAL LANDSCAPE

From the above it is obvious that no such landscape as a single-mono-
cyclic module exists. J. BUDEL (1)   classifies Greece within the belt
of "MIXED SUBTROPICAL REGIONS".

"Subtropical" stands for its inherence to the present day climatic
conditions; "mixed" for all inherited features of the past i.e. since
Plio- and Pleistocene climatic fluctuations.

Evidently, the inevitable question which rises immediatly is the one
which seaks for the starting point. Such starting points may have occurred
more than once during geological times, although it may be assumed
that long periods of geological equability automatically induce a reference level from which new features developed. One may say that it would be
impossible to understand the Quaternary (landscape) if no such reference
level had been developed under the prevailing tropical climatic conditions
of pre-Pliocene periods.

Not only occurred the evolution under long-term equally established
climatic conditions, but the same climate was spread almost all over the
earth from Equator till 70°N. It implies that the reference level is
found over say 80% of the Globe. Moreover, sedimentation remained quite
homogeneous under the prevailing climatic tropical conditions resulting
in broadly developed flat savanna plains with deeply weathered latosols
and isolated inselbergs.The stepwise succession of such plains is due to
tectonical movements lowering the base level of erosion steadily.

The "erosion surface landscape" may in many places be the only upper-
most landscape feature. In mountainous regions, like Greece e.g., such
surface stepped landscape forms the treshold between the lower lying Quaternary scenery and all higher landscapes prior in time to the Tertiary.
In establishing the geomorphologically speaking Teriary-Quaternary boundary one simplifies already the investigation in that the evolution of
the mixed landscape becomes restricted to the lower lying features of the
Quaternary landscape evolution only.

2.2. LANDSCAPE EVOLUTION DURING THE QUATERNARY

Below the Tertiary-Quaternary geomorphological boundary situated
about   200 m. a complex landscape of erosion and aggradation develops
until the present mean sea level (MSL).

An attempt has been made to picturise steps of evolution during the
main climato-sedimentological periods.
(Fig. 1).

2.2.1. LATE OLD PLEISTOCENE (0,8 Ma)

Savanna-marginal tropical climatic conditions have shaped two well
expressed levels at respectively 150 m and 100 m. present day elevation
in most of the Greek scenery.
(Fig. 1/I)

They perfectly reflect the "double denudation process" as described
by J. Büdel (1).

Indeed, latosol development was as before during the Tertiary still
highly effective penetrating duply into the substratum mainly of Mesozoic
Age.

Towards the end of the Early Pleistocene the level at presently 100 m
dominated the landscape comprised a latosol surface interfering with lacustrine basins and lagoons nearly the shoreline.

The surface developed at the beginning of the Early Pleistocene

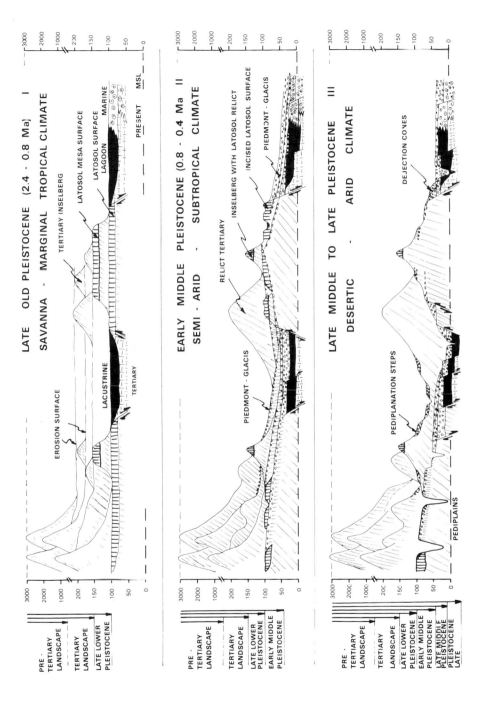

LATE OLD PLEISTOCENE (2.4 - 0.8 Ma) I
SAVANNA - MARGINAL TROPICAL CLIMATE

EARLY MIDDLE PLEISTOCENE (0.8 - 0.4 Ma II
SEMI - ARID - SUBTROPICAL CLIMATE

LATE MIDDLE TO LATE PLEISTOCENE III
DESERTIC - ARID CLIMATE

R. PAEPE, BELGIAN GEOLOGICAL SURVEY, 1984.

sticks out above the 100 m. surface as a latosol covered Mesa surface.
Its platness must have sharply contrasting with the already rounded inselbergs of the preceding Tertiary landscape which usually are covered
with Hughe boulders.

Subsidence in lacustrine and lagoonal area's along block tectonical
faultlines was already active.

Furthermore this mesa-landscape was mainly affected hereafter and
further landscape incision during the Quaternary whereas the former
Tertiary dome form of the inselbergs remained nearly completely intact.
It means that if further evolution did occur later during the Qauternary
it was mainly related to the shaping of the toeslope which became gradually smoother as we shall see hereafter.

It may be questioned why only two such surfaces are generally de-
picted implying two important benchmark stages both with latosol development on top. True enough the entire period of evolution covers certainly 1 Ma years, perhaps 1,5 Ma depending on the initial date of the
start of the Quaternary (J.P. Suc & W.H. Zagwijn) (2).

Anyhow, the deep latosol weathering surface and the independently
developed inselbergs point to a climatical bias of major importance.
As in the lacustrine and lagoonal deposits no such interruption is shown
in the sediment sequence, it may be inferred at tectonic movements in
between the latosol phases.

Should this imply two cycles at 500 K distance in time ? It may be
very possible so as elsewhere e.g. in the N.W. European sequence Tiglian
and Waalian interglacials are equally distanciated in time at 500 K.

In the lacustrine-lagoonal deposits at least 10 cold and warm phases
have been recorded (Fig. 2) implying complete cycles every 100.000 years.

2.2.2. EARLY MIDDLE PLEISTOCENE (0.8-0.4 Ma)

Follows a long period of erosion shaping Quaternary higher mesa's
into pointed isolated hills with latosol relicts on top and a dissected
lower latosol surface embraced by piedmont (glacis) aggradation surfaces (Fig. 1/II).

Tertiary inselbergs disappear greatly or show intensified smooth
toeslope development.

All piedmont fans merge into the subsidence lacustrine or lagoonal
basins where continuous sections of piedmont gravels interrupted by
moderately developed paleosoils are building up the sequence above the
lake / or lagoonal deposits.

In the latter evidence is shown of intensified continuous tectonic
activity which was highly synchronous to the piedmont aggradation itself.

Today they represent hughe series of "brown conglomerates" with an
important series of pedocomplexes (PK) overlying lacustrine or lagoonal
silts of the Early Pleistocene. Another 4 cycles are recorded with pe-
docomplex (PK III, a, b, c, d) ending up each of the conglomerate series.

Clay weathering indexes (J. THOREZ) point at semi-arid, subtropical
climatic conditions, especially during the aggradation of the piedmont
deposits exempt of any vegetation cover (3).

Considering the timespan, also these cycles occurred every 100.000
years, which may be easily compared with the oxygen-isotope stages of
N. Shackleton (4).

2.2.3. LATE MIDDLE TO LATE PLEISTOCENE (0.4 Ma - 10.000 BP)

During the second half of the Middle Pleistocene, while eolian loess

CONTINENTAL STAGES IN GREECE
(COMPARISON O^{18} - STAGES WITH CONTINENTAL STAGES)

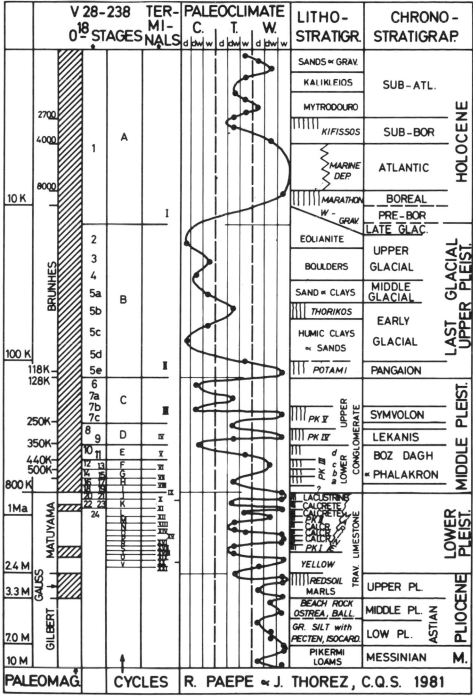

R. PAEPE & J. THOREZ, C.Q.S. 1981

deposits start to cover vast areas of Central and Western Europe, coarse grained eolianites are developed in many parts of Greece. In Macedonia a kind of loessoid sediment is deposited as a vast mantle covering all older piedmont sediments of the earlier part of the Middle Pleistocene. This facies points to a double conclusion : 1) an increasing aridity trend 2) a southern move of the loess belt implying a change of the climato-sedimentological boundary.

Whether one is dealing with the North or the South of Greece, viz. with loessoid or eolianite deposits, the sequence is characterised by a number of fossil soil levels of which two levels, PK IV and PK V are also generally present as broadly developed geomorphological surfaces. Sometimes a doubling (or even tripling) of these surfaces is found which then total up the number of paleosoils in the Middle Pleistocene to at least 9 levels (Fig. 2). Compared to the chronostratigraphic stages of T.A. Wijmstra (5) PK III, PK IV and PK V respectively coincide with BOZ DAGH & PHALAKRON, LEKANIS and SYMVOLON Stages.

During the Late Pleistocene desertification became even more intensive taking into account climatic variance allowing for milder humid interstadial phases.

Apart from the Last Interglacial or Potami Soil (PANGAION SUBSTAGE) no important red soil weathering occurs. Instead one may find steppe like soils such as the Thorikos soil of the Early Last Glacial Stage.

Windborn eolianite deposits seal the sequence at the top and may be dated of the maximum cold about 18000 y.B.P. (Fig. 2).

Within this sequence of soils and eolianites periodicities of roughly 20.000 and 40.000 years are revealed.

As landscape development at the beginning of the Late Pleistocene (Fig. 1/III) started with intensive erosion, eroding deep gullies into piedmont and older deposits, elements of the previous periods were greatly destroyed. This downcutting progressed gradually shaping terrace steps along the inselbergs with piedmont relicts. Aggradation of the younger Late Pleistocene deposits took place in these gullies except for the eolianites which may cover the lower part of the newly established terrace steps. The latter most probably underwent locally the effect of pediplanation processes so that pediplanation steps may occur instead of terrace steps.

In conclusion, the Late Pleistocene was a period of downcutting and widening of the relief elements, under arid climatic conditions.

At the same time, tectonic activity went on thus broadening considerably the size of the subsidence basins.

The result was a low lying flat relief sharply contrasting with hyperbolical inselbergs and high mountainranges. This makes Greece so beautiful.

2.3. LANDSCAPE EVOLUTION DURING HOLOCENE TO PRESENT.

Although the Holocene may be considered as the very last interglacial which is still continuing, its effect on Man's History and changing environment, however, is of so great an importance that special attention must be given to it.

R. PAEPE and M.E. HATZIOTIS worked out in the area of Attica (Greece), more specifically in archaeological excavation sites of Academia Platonos in Athens, in the Marathon Plain and in coastal sites the Temple of Artemis in Brauron (E. Attica) a lithostratigraphy dated on basis of archaeological elements (6). C. BAETEMAN parallely studied the marine sequences where D. TSOUCLIDOU studied the relationship between marine and continental deposits in Brauron.

Putting together all evidences after comparative study of all sites combined, the lithostratigraphic record (Fig. 3) revealed in the Haradros Complex of Marathon six Holocene Soils of which respectively the earliest one (MARATHON SOIL, HS1) and the last one (KALLIKLEIOS SOIL, HS6) are the most developed.

With regard to the Neolithic finds, the Marathon Soil most probably developed about 7.000 BC (9.000 BP); the Kallikleios Soil instead was very accurately dated (725 BC $\pm$ 5 y.) thanks to the presence of Geometrical tombs in many sites of the Academia Platonos.

Strikingly HS 3, HS 4 and HS 5 together with relevant fluvial gravel deposits perfectly encompass the three phases of the Helladic Period. Together with the Kallikleios Soil (H.S. 6) they subdivide the Subboreal Substage into four cycles of approximately 500 y. Soil formations in the fluviatile valley system perfectly tally with peat development in the marine sequence of Marathon. Furthermore, in between soil development phases, fluviatile sedimentation rates score the highest values.

In Marathon, however, no soils are found within the timespan of the geological Atlantic Substage coinciding with the Neolithic. Nevertheless H.S.2 and H.S.1 close the fluvial cycles of respectively Boreal and Pre-Boreal Substage inferring a 1.000 y. periodicity.

This sequence was recently completed with a more detailed profile from Academia Platonos (Kratilou section) (Fig. 4). It produced at least 6 other soils in between H.S.2 and H.S.3 namely : H.S.2 a, b, c, d, e, f. Some of these soils were more weakly developed : gley and steppe soils. It points to the fact that weaker climatic oscillations interfered. However, the presence of these soils testify once more of the 500 years periodicity.

By the time of the development of the Kallikleios Soil about 725 y. B.C. all valleys and coastal plains are completely filled up, to the level very near of today's surface.

As to then sedimentation in general slowed down except for the peaks coinciding with the fluviatile phases which point to high sedimentation rates.

In Marathon as well as in Academia Platonos usually a series of five Holocene soils is recorded : H.S.7, H.S.8, H.S.9, H.S.10 and H.S.11. They induce a periodicity of 500 years. Claymineralogy as well as textual evidence furthermore give evidence of four dry cycles respectively towards the end of the Geometric Period (8 Cent. BC), in the Middle and Late Roman Period (2-4 Cent. AD), in the second half of the 12 Cent. AD and today. They reveal a periodicity of 1.000 years.

For the Holocene it may be concluded at three cycles of periodicity :
- 2.500 years : 1) coinciding with the Subatlantic Substage comprising all archeological Periods after 700 BC i.e. after the Geometric Period till present.

2) coinciding with the Subboreal Stage comprising all Helladic and Geometric Periods.

3) coinciding with the Atlantic Substage comprising the Neolithic Periods.
-1.000 and 500 years : coinciding with minor frequencies and surprisingly tallying with shorter cultural Periods.

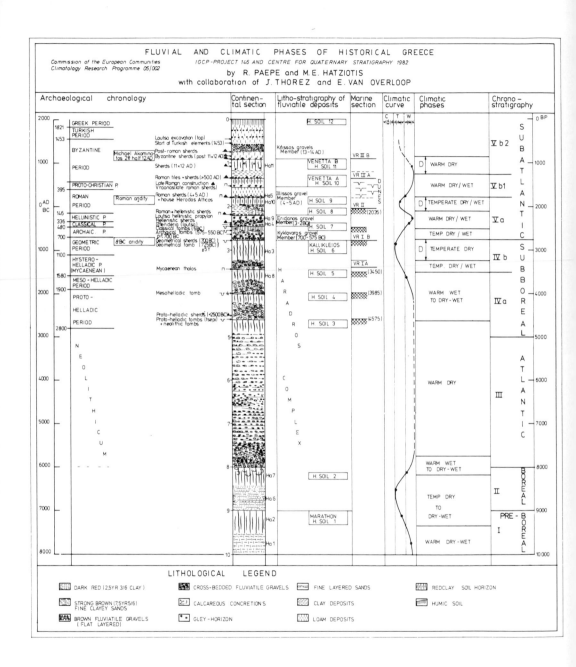

FLUVIAL AND CLIMATIC PHASES OF HISTORICAL GREECE

Commission of the European Communities
Climatology Research Programme 05/002

IGCP-PROJECT 146 AND CENTRE FOR QUATERNARY STRATIGRAPHY 1982

by R. PAEPE and M.E. HATZIOTIS
with collaboration of J. THOREZ and E. VAN OVERLOOP

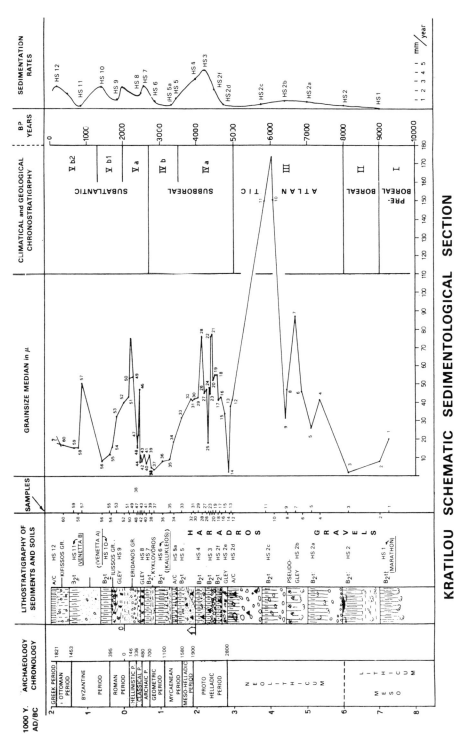

KRATILOU SCHEMATIC SEDIMENTOLOGICAL SECTION

(M.E. HATZIOTIS, R. PAEPE and PAPAY SUPARAN. 1984)

REFERENCES

1. BUDEL, J. (1974). Klima Geomorphologie. Published by Gebrüder Borntraeger Berlin

2. SUC, J.P. and ZAGWIJN, W.H. (1983). Plio-Pleistocene correlations between the northwestern Mediterranean region and northwestern Europe according to recent biostratigraphic and palaeoclimatic data. Boreas, Vol. 12, 153-166. Oslo.

3. THOREZ, J. (1975). Phyllosilicates and clay minerals. pp. 579. Dison. Belgium.

4. SHACKLETON, N. and OPDYKE, N.D. (1976). Oxygen isotope and paleomagnetic stratigraphy of Equatorial Pacific core V28-239, Late Pliocene to Latest Pleistocene. Memoirs geological society of America, 145, 449-64.

5. WIJMSTRA, T.A. (1969). Palynology of the first 30 metres of a 120 m deep section in Northern Greece. Acta Botanica Neerlandica 18. 511-527.

6. PAEPE, R., VAN OVERLOOP, E., HATZIOTIS, M.E., and THOREZ J. (1983). Desertification Cycles in Historical Greece. Progress in Biometeorology, Vol. 3, Swets & Zeitlinger, Lisse.

COMPARISON OF CLIMATIC EVOLUTION DURING POST-GLACIAL TIMES IN GREECE,
TROPICAL AND SUBTROPICAL REGIONS, IN RELATION TO DESERTIFICATION.

E. VAN OVERLOOP
Vrije Universiteit Brussel,
CQS - Centre for Quaternary Stratigraphy
EEC - Project on Climatology

Summary.

The Holocene Epoch is generally, for the Mediterranean basin, scar-
cely subdivided into climatic events. Many literature deals with an
"Older Fill" and a more fluviatile "Younger Fill" (Vita-Finzi).
On the contrary, multidisciplinary study of Holocene sediments in
E. Attica, Greece, yield more than twelve humidity-drought cycles
which occur about each 1.000 years. As from 3.500 BP, the cycles
repeat about each 500 years. (1).
Remarkable is the fact that this evolution is also dealt with in
the lake Tchad area and in Ethiopia. (2)(3)(4)(5)(6)(7)(8)(9).
Some case studies on the historical change in vegetation (10)(11)
and on the failing of damming erosion gullies and reforestation
projects (12)(13) make think about the unavoidable reliance between
an arid climatic phase and desertification, even when induced by
man.

1. COMPARISON OF THE CURVES.

1.1. Sources of the curves. (Table 1)
The table of climatic curves is composed of several compiled data.
The first curve is composed of a tentative part, beneath 3.000 BP,
representing the palynological results of Sabana de Bogotà, Columbia (14),
and of the curve constructed on palynology of the Amazon basin, Brasil
(15).
The second curve is a tentative representation of the interpretation of the palynology on eastern Central Africa lakes. (16)(17).
The third curve shows the most probable climatic oscillations obtained by diatom and lake level studies in Ethiopia. (6)(7)(8)(9).
The fourth curve is taken over from the palynological and lake level
studies in the lake Tchad area. (2)(3)(4)(5).
The curves of northwestern Sahara and the eastern Mediterranean
(Levant + Egypt) are composed of the results of many multidisciplinary
studies. Most of these results came forth from very difficult field ma-
terial, so that their curves are quite incomplete.
The last curve has been established on Holocene sediments in East
Attica, Greece, also by means of multidisciplinary study. (1).

1.2. The eastern mediterranean and N. Africa.

1.2.1. The Late Glacial.
Just before the start of the Late Glacial, the end of the Upper
Pleniglacial (13.000 BP) has generally been marked by an increase of

moisture and/or rainfall.

The area around the Dead Sea points at an increase in rainfall until 12.000 BP. (18). The same occurs in Lebanon - Syria until 13.000 BP (19) after which dry conditions become dominating.

Palynology in Northern Israël (20)(21) points at a recovery of the forest as from 14.000 BP. Hereafter a short dry phase (12.500 - 12.000 BP) seems to have interrupted the moistening in the Levant. (19)(22)(18).

Optimal climatologic conditions prevail between 12.000 and 11.000 BP as shown by pollen analysis. (20)(21). This period coincides with the flourishing of the Natufian culture in the Levant and Syria.

The Nile area and Egypt give evidence of decreasing floods and local winter rainfalls, which last from 17.000 BP until 8.500 BP. A dry interstade occurs at about 11.500 BP. (23).

For other regions, except for Greece, no precise datable information exists except for some broad statements about the Late Glacial as a whole.

The Sahara and North Africa also show a generally moister and even warmer climate however, with more local rainfall and locally dryer areas. (18).

Humidity is furthermore better expressed in the Central Sahara than in its northern part. (18).

In Iran, the existing Artemisia-steppe slightly changes into a beginning Quercus-Pistacia woodland, which points again at increasing moisture. (20)(21). Geomorphology in Iran shows brief and very violent rainfall for this period. The relatively low temperatures are responsible for the increasing moisture.

The situation in Greece was almost the same; herbs slowly left their places for trees (Quercus-Pistacia), and here too an increase of moisture was due to low evapotranspiration. (20)(21). Turkey shows a long lasting drought due to higher temperatures. (20)(21).
The post-Glacial therefore may generally be identified by a slight amelioration of the climatic conditions due to an increase of humidity.

1.2.2. Preboreal and Boreal (10.500 - 7.500 BP).

After the drier period in the Levant (11.000 - 9.000 BP), pollen analyses, archaeology, and geomorphology indicate again wetter environment during the 10th millennium BP.
The climate is even moister than at present. (18)(22)(25). From 9.000 BP on, one meets with an increasing drought with a climax of aridity around 8.500 BP.

The same is attested in Syria and Lebanon, where palynology on settlements (26)(21)(18) shows a steady drop in arboreal pollen percentage.

In Iran and Turkey (27)(20)(21), the inner area gradually is invaded by a forest-steppe during this timespan.

On the contrary Greece already is covered by a dense oak forest (20)(21) i.e. the climax vegetation, under increasing temperatures and dry environmental conditions, which turn to an increase in precipitation towards the end of the Boreal (8.000 BP).
At this very moment too, R. Paepe and M. Hatziotis locate the Marathon Soil, coinciding with the beginning of the Neolithicum, and as determined by J. Thorez on basis of clay mineralogy testifying of warmer and more humid conditions as well. (1).

As a general conclusion the eastern part of the Mediterranean has been favoured by a moistening of the climate during the Preboreal, with a steady throughout the Boreal followed by a humidity-peak around 8.000

BP. (1)(Table 1).
    In the Central Sahara, increased rainfall lasts until 9.000 - 8.000
BP interrupted by dry phases. (18).
For the Northern Sahara, a wet phase is known to occur between 8.500 -
7.000 BP. (18).
In Egypt, the local rainfall remains up to 8.500 BP. (23).
    The southern part of the Mediterranean testifies of regional va-
riance of rainfall-dispersion, but the general tendency as described
for the eastern part of the Mediterranean.

1.2.3. Atlantic (8) (7.500 - 5.500 BP).
    The Atlantic is known as being a period of intensive human settle-
ment in all regions.
    In the Levant, settlement during the first part of the Atlantic re-
mains in isolated areas whereas the most intensive period of dispersion
of sites occurs between 6.000 and 5.500 BP as a result of greater agriculture possibilities due to higher humidity. (18)(28).
Also pollen analysis in this region points at warmer and moister climatic conditions during the Atlantic. (29). Furthermore the Dead Sea clearly shows a higher level at the end of the Atlantic (5.500 - 4.300 BP),
due to more humid conditions.
    Northern Israël climate between 7.500 and 5.500 BP is moister and
comparable to the today's one. The pollen evolution is comparable to the
Western European one (29), i.e. the installation of the climax vegetation.
    Syria and Lebanon also was dealing with climatic conditions during
the Atlantic period (26) similar of today. A highly moist phase at 5.000
BP is shown, whereas geomorphologic evidence points at higher rainfall
between 7.500 - 5.800 BP. (18).
    In Arabia, investigation concerning the Atlantic reveals pluvial
periods till about 6.000 BP. (30).
    North African climate in the Atlantic period first crosses an ari-
dity phase, followed by a period of higher moisture between 6.500/6.000
and 5.000 BP. (18).
Lakes develop in the western Sahara from 8.000 till 4.000 BP, due to optimal climatic conditions (31) as well.
Iranian palynological investigations yield the final establishment of
the Quercus climax forest at the end of the Atlantic period, favoured
by a warm and wet climate. (27).
    Evenso in Turkey, the climax vegetation comes to growth, also in
mountaneous areas.
    In Northern Greece, former palynological studies (24)(20)(21) give
prove of the final installation of the climax deciduous oak forest at
the end of the Atlantic, as a result of an increas in temperature and
humidity during this period. The archaeogeological evidences collected
by R. Paepe, M. Hatziotis, J. Thorez and E. Van Overloop (1), in East-
Attica (Greece), point at warm and intermittent dry climatic conditions
during the Atlantic, a tendency which lasts until 3.900 BP, i.e. the end
of the Protohelladic period coinciding with the middle of the Subboreal.
For this period, four dry-wet cycles are up till now discovered (M. Hatziotis, oral communication).
    The general trend during the Atlantic is in every part of the Me-
diterranean one of continuous moistening, with a humidity peak differing
in time from area to area.
The intermittent dry-wet situation in East Attica (1) makes us believe of

more existing oscillations during the Atlantic in all regions concerned, which are to be envisaged by deeper investigations on soils, pollen and archaeology on more detailed sections, if available.

1.2.4. The Subboreal (5.500 - 2.700 BP).
Results obtained from disciplines as varied as pollen analysis, geomorphology and archeology point at more humid and cool climate for the beginning of the Subboreal (5.500 - ± 4.000 BP) (28)(32)(18) in the Levant.
Syrian palynology (20)(21) gives prove of a semi-arid climate for the Subboreal.
Pias et al. (18) show the existence in the NW part of the Sahara of a cooler and more humid climate about the transition from Atlantic to Subboreal changing into warm and dry one towards the end of Subboreal.
Palynological investigation in Turkey (20)(21) prove the definitive installation of the climax forest coinciding with further moistening of the climate due to lower evapotranspiration as a result of lower temperature.
Also a NW Greece (20)(21) the already existing climax woodland remains allmost unchanged, despite cooling and moistening of the climate. Bintliff (33) concludes that climatic changes are only observed in the so-called "marginal" woodlands i.e. the ones growing in temporarily favoured areas where desertification is able to extend very rapidly due to local environmental factors.
Iraq deals for the same period with two big flood-periods (34)(35) respectively dated at 4.850 - 4.300 BP and 3.800 - 3.450 BP. The last mentioned flood coincides with a humid period in Greece. (1).
In Eastern Attica (1) these floods are also represented by the upper part of the Haradros Gravel complex, a series of continental gravels and/or peat formations in the coastal plain of Marathon. They correspond to three cycles of sedimentation each of which is coinciding with respectively the Proto-Helladic, Meso-Helladic and Hystero-Helladic (Mycaenian) Periods separated by a series of Holocene Soils : H.S.3, H.S.4, H.S.5. A change from warm-wet to temperate dry climate at the end of the Subboreal is found, a drought which probably is at the origin of the fall of the Mycaenian culture. (18). Then dry conditions enchanced the development of the "Kallikleios" Soil, H.S.6. (1).
The Subboreal is thus clearly dominated by a change in climate which has marked almost every area of the Mediterranean in a similar way : i.e. cooling-off and the transition from wet to dry climate, especially in Greece, resulting in a soil development.

1.2.5. The Subatlantic (2.700 BP till Present).
During this period the climate in the Levant becomes almost the same as the present one, although in some cases less extreme. The Sinai and surrounding area's show evidence for settlement from ± 2.800 BP until ± 2.000 BP (18), probably as result of increased humidity favouring agriculture which is responsible for the human occupation.
Moreover, stratigraphic evidence in the Levant witnesses of a "Younger Fill", which represents very thick fluvial deposits, dated by Vita-Finzi (18) between 1.600 and 200 BP or by Goldberg (22) between 1.700 and 600 BP.
The "Younger Fill" is also dealt with in Syria at the very same period as in the Levant (18), and further on in Iran (18) and the Gulf (Diester-Haass, 1973) where fluviatile "Younger Fill" deposits seem interrupted by dry intervals.

SYNOPTIC TABLE OF POSTGLACIAL CLIMATIC CHANGES (I).

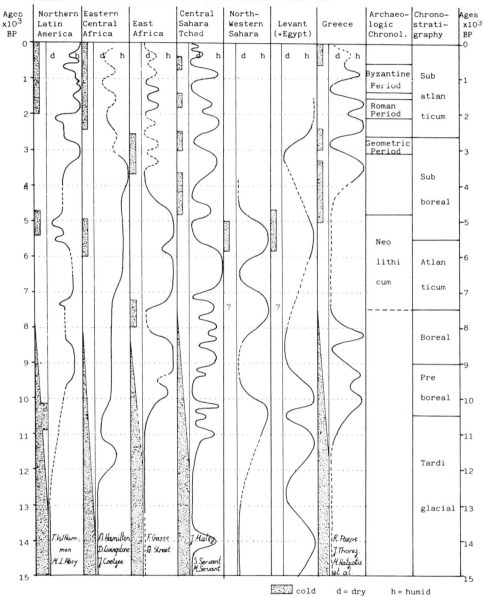

by Elfi VAN OVERLOOP (1984)

FLUVIAL AND CLIMATIC PHASES OF HISTORICAL GREECE

Commission of the European Communities
Climatology Research Programme 05/002

IGCP-PROJECT 146 AND CENTRE FOR QUATERNARY STRATIGRAPHY 1982

by R. PAEPE and M.E. HATZIOTIS
with collaboration of J. THOREZ and E. VAN OVERLOOP

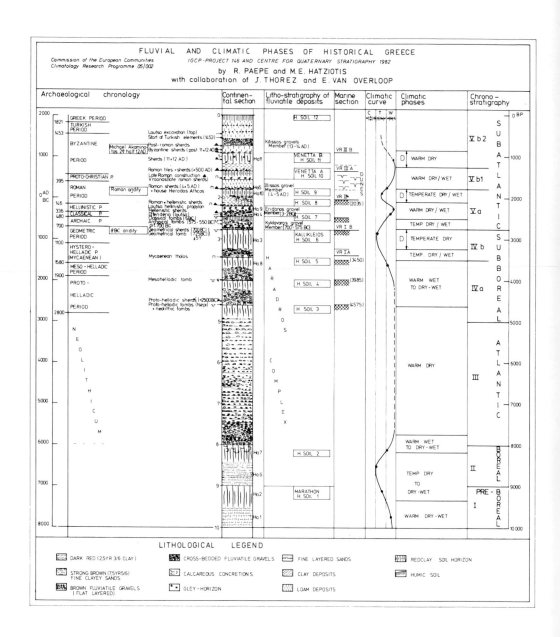

In Turkey, the "Younger Fill" is largely represented as well (18) dated however from ± 1.600 BP till Present.

As for the Subboreal, pollen-analytical diagrams show little if any change in their pattern during the Subatlantic. The climate changes were indeed too short to give general modifications in the wood cover, whereas from 3.000 BP on, human influence plays an important role in the vegetation assemblage.

The same vegetational historical problem exists for Greece (20)(21) (24) where too small time intervals of climatic changes and too intensive human occupation occured. Anyhow, Bottema (20)(21) mentiones cooler and relatively moist Subatlantic conditions in Greece.

It has been possible thanks to transdisciplinary research between archeology, stratigraphy, clay mineralogy and sedimentology to subdivide the Subatlantic in even more detailed climatologic events. (1)(Table 2).

Four main fluviatile gravel-beds are identified belonging to the last 2.700 years, each time coinciding with a change of human cultures; Kyklovoros Gravels (700 - 575 BC), coinciding mainly with the Archaic Period and resting on the higher mentioned Kallikleios Soil (H.S.6); Eridanos Gravels (3 - 2 Century BC) comprising Classical and Hellenistic Periods overlying H.S.7; Ilissos Gravels (4 - 5 Century AD) overlying II.S.8 and 9 and finally Kifissos Gravels (13 - 14 AD) overlying Holocene Soils 10 and 11 i.e. Venetta Soils. These evidences allow to subdivide the Subatlantic into five climatologic divisions (1) namely a temperature dry-wet phase, a warm dry-wet phase, followed by another temperature dry-wet phase of the end of the Roman Period; a warm dry-wet one coinciding with the Proto-Christian and the beginning of the Byzantine Period marked by the Venetta Soil development. The middle of the Byzantine Period is marked by a warm dry environment which is also mentioned by Michael Akominatos (2nd half 12 AD).
The sixth climatic subdivision shows a jump into the warm and wet conditions lasting during the Ottoman Period, characterised by sandy gravel deposits, until the Present-day conditions.

Thus, the so-called "Younger Fill" deposits comprise only the Ilissos and Kifissos - Gravels of the detailed subdivision of Paepe, Hatziotis, Thorez and Van Overloop (1982) (1). They incorporate three desertification periods i.e. the Roman, the Byzantine and the Present-Day Periods.

1.3. Central Sahara-Tchad.

Diatom studies and palynological investigations on the lacustrine sediments of lake Tchad (4)(5)(2)(3) provided the scientific world with some of the most detailed information on climatic fluctuations in the lake Tchad basin. After compilation of all the data of these works, the Late Glacial and the Holocene seem to be composed of not less than seventeen humidity-drought cycles, the eldest one occuring at ± 14.000 BP.

Between 11.000 BP and 10.000 BP, three cycles are noticed with a peak in humidity at the base of the Preboreal.

An 800 years drought persists around 9.700 BP whereafter five humidity peaks have been discovered up until the boundary between the Boreal and the Atlantic (7.500 BP).

The very Atlantic can be compared to the "climatic optimum" of the Western Sahara (31) since a very wet and warm climate coincides with this Period.

As from 5.000 BP on till recent, seven humid dry cycles occur.
Remarkable is the similarity between the course of the cycles for
Greece and for the Tchad region, especially during the Subatlantic.
Although lake Tchad ranges under the so-called "marginal"climatic regions and one could therefore expect to some extent detailed investigation results, then similarity with Greece again proves that precise
results can also be obtained in less "marginal" or sensible regions of
the earth, by means of collaboration between several disciplines.

1.4. Central African and eastern African lakes.

Central African and eastern African lakes yield detailed studies on
palynology from high mountaineous lakes in the western part of the great
Rift lakes and on diatoms from eastern Rift lakes. The palynological
diagram from lake Mahoma, mount Ruwenzori, Uganda, elaborated by
Livingstone (1967) and reinterpreted by Hamilton (1972), (16) together
with the general overlook on palynological data from several mountains
in the same region (17) have been selected in order to establish a
temptative construction of a climatic curve considering the changes in
humidity and temperature in Central Africa. As for East Africa, the same
procedure has been followed, using the data from diatom studies on the
lakes of the Afar-region, Ethiopia. (6)(7)(8)(9).

1.4.1. Central African lakes.
The curve on summarized data which are described above (16)(17)
yields a very dry Late Glacial, colder than today. In comparison with
the other discussed regions, the upwarming and mountaineous deglaciation in Eastern Central Africa takes place very early, namely as from
± 14.000 BP on. A phase of increased humidity is situated at about
11.500 BP. The today's climatic conditions are established quite directly
after this upwarming.
A more humid phase is mentioned lasting during the Preboreal and the
Boreal, changing into a period of still more humidity which coincides
with the end of the Atlantic and the beginning of the Subboreal, this
last one accompanied by a lowering of the temperature.
The very "climatic opticum" is for eastern Central Africa situated
in the second half of the Subboreal, with high humidity and temperatures.
The Subatlantic so far has not been subdivided into detailed oscillations of drought and humidity. As for the pollendiagram of Livingstone
(1967), reinterpreted by Hamilton (1972) (16), a possible subdivision
seems to proceed out of the curves of certain well specified taxa of
humidity-related plants, of which the pollen sediment locally or over
long distances (Personal interpretation). Remarkable is the fact that
these curves do proceed or recede at nearly the same time, and that the
curve of the taxon indicating "human influence" behalves quite independantly of the curves of the humidity indicators. Hence the human influence on these moist-indicating taxa seems to have been of less importance.
Four dryer periods seems to occur as from the end of the Subboreal
(2.700 BP), interrupted by periods of increased humidity, each oscillation covering about 500 years.
It would be of very much interest to put the palynology of the
Subatlantic on the Central African mountains into more detailed investigation, as the features of human disturbance are well known in botany
and did in historical times not affect the whole natural, especially the
mountaineous vegetation.

1.4.2. Eastern Africa.
F. Gasse and E. Street (6)(7)(8)(9) worked out, by means of diatom
studies. several detailed oscillation curves of lake levels in the high
mountains of Ethiopia.
F. Gasse evenmore points out, especially for the Tardiglacial and
the Holocene, which were the specific palaeoenvironment in which the
diatoms were living. Compilling these data, a temptative climatic curve
has been elaborated.
The deglaciation took place before 11.500 BP, since at this very
date, the present environmental conditions were already established.
Four levels of high humidity are recognized during the Lower and
Middle Holocene (up until 3.500 BP); the first one occurs at ± 10.000 BP.
The second one covers the end of the Preboreal and the Beginning of the
Boreal (up till ± 8.200 BP). The third one starts with the Atlantic, after a dry and colder oscillation, and ends with a short dryer phase
around 6.000 BP. The fourth one coincides with the base of the Subboreal,
after which a longer drought lasts (from ± 4.500 BP - ± 3.500 BP) at the
end of which a cooling-off is registered.
As from 3.500 BP on, F. Gasse describes an hypothetic cyclicity of
450 - 500 years, expressed by the alternation of wet alcaline and dry
acid environmental lacustrine conditions (6), cyclicity which has been
confirmed by the diatom and lake level studies on other lakes of lower
altitude (6)(7)(8)(9), probably pointing at a cyclicity in local rainfall pattern.
For the whole Holocene, an eleven-cycles curve of dry-wet situations is
suggested, of which four occur in the Lower and Middle Holocene and
eight in the Upper Holocene.
As for the results from Lake Tchad area, there again the similarity
with the Greek curves is striking, especially when looking at the number
of cycles and their timespan.

1.5. Northern South America.

Both the palynological investigations on the "Sabana de Bogota",
situated on the high plateau's of Columbia (14) and the holocene palyno-
logic investigations in the Amazone basin, Brasil (15)
yield sufficient evidence for the construction of a tentative curve for
the Lower and Middle Holocene to which the curve for the Upper Holocene
from the work of M.L. Absy (15), is added.
The Tardiglacial is again very dry in northern South America and the
upwarming starts at about 13.000 BP, with a new cooling-off between
11.000 BP and 10.000 BP.
A humid phase covers the end of the Preboreal and the beginning of
the Borea (9.000 BP). Moreover, this humid phase is also found back in
Minas Gerais (Brazil), by means of palynological investigation on a
lacustrine peat bog, indicating a transition from high mountaineous vegetation towards the installation of the atlantic-tropical rainforest.
This transition coincides with a high lake-level phase around 9.000 BP.
(11).
At the boundary between the Boreal and the Atlantic, a short dryer
phase is recorded, following a period of intermediate wet-dry climate.
High temperature and humidity characterise the Atlantic.
Two dry oscillations occur at the base of the Subboreal (± 5.500 BP
and ± 5.000 BP) accompanied by a lowering of the temperature.
As from 3.000 BP on, three main cycles of higher humidity are dis-

covered, interrupted by dryer phases. The second cycle shows three peak's and three periods of somewhat dryer conditions.

The first dry phase occurs from 2.700 BP up until ± 2.000 BP, after having passed through a maximum drought around 2.100 BP, next areas are situated at ± 1.000 BP, and are much less expressed.
Moreover, two dry periods occur at ± 700 BP and ± 400 BP (15); the last dry period coinciding with the Western European "Little Ice Age".

1.5.1. Discussion.

Great similarities exist between the curves from Tchad area and Greece. Both of them are as well comparable to the curve from East Africa.
On the other hand, the curves from nothern South America and from eastern Central Africa show many coincidences, which are on themselves again comparable to the East African curve.

Since climatic data from the Levant area and Egypt are, taking into concern the difficult field material, not very abundant, the curve from these regions does not appear sufficiently detailed as to compare it to other curves. Unless this fact, some of its events may be found back in the Greek curve as well.

The curve which behaves quite independantly is the one from the Atlas Mountains, say northwestern Sahara.

The deglaciation during the Late Glacial took generally place in very early times in tropical regions (before 13.000 BP) and appearingly much later in the subtropical regions (from 12.000 BP on).

Remarkable is the fact that the period of the so-called" climatic optimum", in some regions takes place during the Atlantic (7.500 BP - 5.500 BP), in other regions (i.e. eastern Central Africa and north-western Sahara) during Subboreal (5.500 BP - 2.700 BP).

Whereas the minor climatic oscillations during the "climatic optimum" in the other regions are not yet found back, Greece yields evidence of a four-fold entermittent dry-wet climate for that period.

Whilst the most detailed curves (Greece and Tchad) yield at least fifteen dry-wet cycles, the curves from northern Latin America and East Africa at least procure eight of them (hypothetically eleven in East Africa).

Sedimentological investigation (Paepe & Suparan, 1984, in press) would even yield not less than nineteen cycles for E. Attica, Greece.

Many authors discuss about the fact that the Lower and Middle Holocene would cover relatively less oscillations than the Upper Holocene. Many times, for the Mediterranean basin, only an "Older Fill" and "Younger Fill" is mentioned, without any precise subdivision or climatic interpretation.

On the contrary, almost all the discussed curves show a major oscillation of 1.000 - 1.500 - 2.000 years, which generally becomes more detailed in the Subatlantic (as from 2.700 BP), i.e. with a 450 - 500 years rate.

Looking at the curve of lake Tchad, even the Lower and Middle Holocene are subdivided into about 500 years oscillations.

The results of sedimentological study in E. Attica, Greece, are now pointing at a similar evolution (Paepe & Suparan, in press).

Taking these criteria into account, an arid climatic phase repeats each 500 years, of which the last one is a present in the Mediterranean area and the Subtropics.

2. EXAMPLES OF CLIMATE-RELATED HISTORICAL VEGETATION EVOLUTION.

2.1. The Tropics : Rwanda (10).

The here mentioned study has been carried out on iron-melting fires of the Iron Age on a high-plain, 1.5 km south of Butare (Rwanda). Palynology and dating of the fires made it possible to find out the movements of the population and the related evolution of the vegetation.

Roche and Van Grunderbeek found an arid phase around 2.685 BP, the one which also occurs in E. Attica (Greece), and is noticed over the world.

At the date of 1.775 BP, around the site where people lived at that time, the area was covered by a dense park-savanah. However, at 1.600 BP, the forest here was degraded and became an open park-savanah, because of burning of the wood in the fires. Iron-Age men left the area to go and live more to the West.

The vegetation recovered the abandoned area between 1.600 BP and ± 1.400 BP.

When looking at the climatic evolution (Table 1) between 1.800 BP and 1.200 BP, the climate was favourable for natural reforestation. No aridity is dealt with at that moment. Hence, after abandonment of the environment, natural vegetation is able to come back.

2.2. Greece : Pollen results from Roman times (36).

Samples from sediments of E. Attica (Greece), dating from the Roman Period (2.100 BP - 1.600 BP) were examined on pollen content. The precise datation has been determined by M. Hatziotis on archaeological base. Hence, the region was occupied at that time.

Going up in time, there is a clear decline in tree pollen (especially Pine, Rosaceae trees, Oleaceae) in favour of "steppe pollen" (especially Gramineae and Artemisia). Furthermore, species typical for cultivation almost disappear.

Appearingly, the environment went drier, since cultivation stopped. The forested area became a forest-steppe. This aridification of the vegetation is possibly due to human intervention, although cultivated species disappear, which makes think that occupation went on elsewhere.

Moreover, no signs of recovery of the vegetation are seen in the pollen content.

Taking into account the climate curve from E. Attica, Greece (1) a dry phase occured around 1.800 BP, in the middle of the Roman Period.

Due to this dry period, the vegetation was most probably not able to recover in a natural way.

3. CONCLUSION.

The evidence exists all over the world of an alternation during the Holocene of dry and humid climatic phases. Especially from 2.700 BP on, these cycles return each 450 - 500 years. One can expect an aridification of the climate showing the same cyclicity.

It can be stated that when climate is favourable (humid phases in climatic cycle) an anthropogenic desert (woodland to Savannah or steppe) can recover in a natural way.

On the contrary, when climate turns into an arid phase, natural recovery is not longer possible, and vegetation will either keep a desertic aspect, either disappear, when soils are eroded. Many regions of the Mediterranean nowadays deal with such an arid phase. In the case that soils are still present, reforestation is possible (Primbe, Votanicos Kipos, Athinai, oral communication).

In the case of complete desertification, this means huge soil erosion (37), the species used for reforestation should not only be adapted at the subsoil (13) but also at the climate (dry species, steppe species), whereafter shrubs and small trees can be tried out.

Attention is stressed on the fact that an arid climatic phase induces environmental conditions asking for an adapted treatment (botanical or engineering). Many failed projects in this respect (12) prove that not enough attention has been made to the climatic cyclic character of aridifications. A small arid phase lasts for several hundreds of years and will in the future repeat with the same cyclicity as in historical times.

In this respect, all environmental projects of the future have to envisage climatic evolutions.

REFERENCES.

1. PAEPE, R., HATZIOTIS, M.E., VAN OVERLOOP, E., THOREZ, J. (1982) Desertification cycles in Historical Greece. EEC workshop on palaeoclimatology, ed. M. GHAZI
2. MALEY, J. (1977). Paleoclimates of Central Sahara during the early Holocene. Nature, vol. 267, nr. 5629, 573-577, Macmillan Journals, Ltd.
3. MALEY, J. (1977). Analyses polliniques et paléoclimatologie des douze derniers millénaires du bassin du Tchad (Afrique centrale). Recherches Françaises sur le Quaternaire, INQUA. Suppl. au Bulletin AFEQ, 1977-1, nr. 50.
4. SERVANT, S. (1970). Répartition des diatomées dans les séquences lacustres holocènes au Nord-est du lac Tchad. Cah. ORSTOM, sér. Géol., Paris, 2(1), 115-126.
5. SERVANT, M.; SERVANT, S.; LARMOUZE, J.-P., FONTES, J.-C.; MALEY, J. (1976). Paléolimnologie des lacs du Quaternaire récent du bassin du Tchad. Interprétations paléoclimatologiques. IInd Int. Symp. Paléolimn., Pologne, 23 p.
6. GASSE, F. (1978). Les diatomées Holocènes d'une tourbière (4.040 m) d'une montagne Ethiopienne : Le mont Badda; Rev. Algol., M.S. XIII, 2; 105-149
7. GASSE, F. & STREET, F.A. Late Quaternary lake level fluctuations and environments of the Northern Rift valley and Afar region (Ethiopia & Djibouti).
    (1978). Paleogeography, - climatology, - ecology, 24 : 279-325, Elsevier Sc. publ. Cy., Amsterdam.
8. GASSE, F. & DELIBRIAS, G. Les lacs de l'Afar central (Ethiopie et TFAI) au Pleistocène Supérieur.
    (1976). Paleoclimatology of Lake Biwa & the Japanese Pleistocene, vol. 4 (ed. S. Horie).
9. GASSE, F. (1977). Evolution of Lake Abhé (Ethiopia & TFAI), from 70.000 BP. Nature, vol. 265, nr. 5589, 42-45, Macmillan Journals, Ltd.
10. ROCHE, E. & VAN GRUNDERBEEK (1984). Le premier Age du Fer au Rwanda et au Burundi. Archéologie et environnement I.N.P.S., Butare, Rwanda, publ. 21.

11. VAN OVERLOOP, E. (1981). Postglacial to Holocene transition in a peatlayer of Lake Jacaré (Rio Doce basin, Brasil). Bull. Soc. Belge Géol., T. 90, fasc. 2, 107-119.

12. RENDELL, H. (1986). Soil erosion and land degradation in Southern Italy. Int. Sem. Desert. Europe - Mytilini, Greece, CEE.

13. MENSCHING, H.G. (1986). Processes of desertification in Southern Europe - Examples (Sicily - Spain). Int. Sem. Desert. Europe - Mytilini, Greece, CEE.

14. VAN DER HAMMEN, T. Historia de clima y vegetacion del Pleistoceno Superior y del Holoceno de la Sabana de Bogota. Bol. Geol., vol. XI, nr. 1-3, 189-266

15. ABSY, M.L. (1979). A palinological study of Holocene sediments in the Amazon basin. Doctoraal Proefschrift, V.U. Amsterdam, Nederland.

16. HAMILTON, M.A. (1972). The interpretation of pollen diagrams from Highland Uganda. Palaeoec. of Africa, vol. 13, 45-150.

17. VAN ZINDEREN-BAKKER, E.M. Sr. & COETZEE, J.A. (1980). Are-appraisal of Late Quaternary climatic evidence from tropical Africa. Palaeo-ecology of Africa, vol. 13, 151-181.

18. ROGNON, P. (1980). Interprétation paléoclimatique des changements d'environnements en Afrique du Nord et au Moyen Orient durant les 20 derniers millénaire. Coll. CNRS, nr. 598, Lyon.

19. LEROI-GOURHAN, A. (1980). Diagrammes polliniques de sites archéolo-giques au Moyen Orient. Coll. CNRS, N° 598, Lyon.

20. VAN ZEIST, W. & BOTTEMA, S. (1980). Palynological evidence for the climatic history of the Near East, 50.000 - 6.000 BP. Coll. CNRS N° 598, Lyon.

21. VAN ZEIST, W. & BOTTEMA, S. (1980). Vegetation history of the Eastern Mediterranean and the Near East during the last 20.000 years. Symp. Env. East. Med., N. East, Gron.

22. GOLDBERG, P. & BAR-YOUSSEF, O.(1980). Environmental and archaeolo-gical evidence for climatic change in the southern Levant and ad-jacent areas. Symp. Env. East. Med., N. East, Groningen.

23. BUTZER, K.W. (1975). Patterns of environmental changes in the Near East during Late Pleistocene and Early Holocene times. In : Probl. Prehist. : North Africa & Levant, South. Neth. Univ. Dallas, 389-410.

24. WIJMSTRA, T.A. (1969). Palynology in northern Greece. Acta Bot. Neerl. (18)4.

25. TCHERNOV, E. (1980). Faunal responses to environmental changes in the eastern Mediterranean during the last 20.000 years. Symp. Env. East. Med., N. East, Groningen.

26. LEROI-GOURHAN, A. (1974). Etudes palynologiques des derniers 11.000 ans en Syrie semi-désertique. Paléorient 2|2, 443-451.

27. VAN ZEIST, W. & BOTTEMA, S. (1977). Palynological investigations in Western Iran. Palaeohistoria 19, 19-85.

28. PRICE-WILLIAMS, D. (1975). The environmental background of prehisto-ric sites in the Fara region of the western Negev. Bull. Inst. Ar-cheol. 12, 125-143.

29. HOROWITZ, A. (1971). Climatic and vegetational development in north-eastern Israël during Upper Pleistocene-Holocene times. Pollen & Spores, 13, 255-278.

30. Mc. CLURE, H.A. (1976). Radiocarbon chronology of Late Quaternary lakes in the Arabian desert. Nature 263, 755-756.

31. PETIT-MAIRE, N. (1977). Congresso Brasileiro de formaçaos super-

ficiais, S.P., Brasil.

32. OATES. J. (1980). Prehistoric settlements patterns in the Middle East in relation to environmental conditions. Symp. Env. East. Med., N. East, Groningen.

33. BINTLIFF, J.L. (1980). Palaeoclimatic interpretation of environmental change. Symp. Env. East. Med., N. East, Groningen.

34. PAEPE, R.; GASCHE, H.; DE MEYER, L. (1978). The surrounding wall of Tell ed-Dèr in relation to the regional fluviatile system. Tell ed-Dèr II, Leuven, rep. nr. I, IGCP 146, 1-35.

35. PAEPE, R.; BAETEMAN, C. (1978). The fluvial system between Tell ed-Dèr and Tell Abu Habbah. Tell ed-Dèr II, Leuven, nr. 2, IGCP 146, 37-56.

36. VAN OVERLOOP, E. (1982). Pollenanalysis on some archeologic sites in E. Attica, Greece. Rep. EEC progr. Climat.

37. MANCINI, F. (1986). Soil Conservation problems in Italy after the Council of Research finalized project. Int. Sem. Desert. Europe-Mytilini, Greece, CEE.

CLIMATIC IMPLICATIONS OF GLACIER FLUCTUATIONS

J. M. GROVE
Girton College, Cambridge.

Summary

For over a century glaciers all over the world have been retreating
in response to rising temperatures, especially summer temperatures.
Retreat has been punctuated by advances at intervals of a few decades
which appear to be synchronous in many widely separated mountain
regions. Glacier fluctuations and the moraine and other deposits
they leave behind them provide opportunities to trace the climatic
record well before the instrumental period and thus provide a context
for current climatic variability. Small scale climatic variations
particularly affect oceanic highland areas, where the growing season
depends heavily on summer precipitation. The association between
temperature fluctuations in north-western Europe and precipitation
fluctuations in southern Europe requires further investigation.

European glaciers, it is well known, have retreated quite drastically
since the middle of the last century, in the cases of the largest by a
kilometre or more. They have also thinned, leaving mountain climbing huts,
sited conveniently in the nineteenth century, perched high above the ice.
It is less well-known that retreat has been discontinuous and that the
interruptions have been nearly synchronous across Alpine Europe. Since
1850, minor readvances affecting the majority of the glaciers have been
widespread, about 1890, 1920 and since 1960 (Figure 1). More surprisingly,
minor readvances with similar or identical timing have been recognised in
other parts of the world where conditions are very different from those in
Alpine and Scandinavian Europe. During the last two decades glaciers have
enlarged in Alaska and Arctic Canada, the Caucasus, Tibet and Nepal, Peru,
New Zealand and some of the Sub-Antarctic islands (Figure 2).

General glacial retreat has occurred as the level of the snowline has
risen by 100 to 300 m., and mean annual temperature has increased by about
1 C. Such a temperature change has been measured instrumentally in Europe
and elsewhere (1, 2). Temperature is not the sole control of glacier
volume and extent. Precipitation, especially in the form of snow is also
important and the response of a particular glacier to a climatic fluctuation is affected by its altitudinal distribution, shape, aspect and the
presence or absence of a heavy debris cover. Nevertheless it is apparent
that summer temperature affecting the length and intensity of the ablation
season is the dominant climatic control. If a large sample of glaciers is
considered, the general pattern of oscillations becomes quite clear.

A great deal of historical information is available about European
glaciers, as their fluctuations have had important consequences for those
living and working near them and also for travellers using high mountain
routes. The importance of past glacier volume and extent as a climatic
record was recognised in Switzerland by the hydrological engineer Venetz
in the 1830's (3) and the first systematic project to measure the positions
of glacier tongues was initiated in the 1880's. Similar arrangements were
made in Austria and France in the following decades and in Italy somewhat
later. Consequently we now have a substantial body of quantitative data

to correlate with the European meteorological records. This, together
with modern research on meteorological controls of glacier mass balance
(4), provides a basis for examining the historic record of glacier fluctuations and its meteorological implications. Support is provided by studies
of tree-ring width and density variations, for tree growth is also
dominantly determined by summer temperatures over much of Europe (5,6).
Both sets of investigations reveal that in the late sixteenth century
climatic conditions in Europe became swiftly harsher with shorter, cooler
summers and harder winters. Snowlines were lowered by 100 to 200m. and
glaciers advanced in many places onto agricultural land. Mean annual temperatures were 1 or 2C. lower. This was the cooling from which we have been
recovering since the middle of the last century. Between 1600 and 1850 conditions were by no means constant. Though the glaciers never withdrew to
the positions they occupied in medieval times, or occupied again today,
their fronts oscillated as temperature and precipitation varied.
        It is now apparent that the same kind of thing was happening in other
continents. For the present century we have a good deal of observational
data and actual measurements from some regions. For earlier centuries the
sequence of events has to be pieced together by dating the moraines formed
in phases of glacial expansion or standstill. Dating can be accurate to
the nearest 2 years for the nineteenth century (7) but usually the accuracy
is less and uncertainty increases as we penetrate further into the past.
Many of the ages given are accurate to only $\pm20\%$ at best (8). If all the
available data is plotted by latitude it becomes quite clear that the
climatic deterioration of the period that has become known as the Little
Ice Age was a global phenomenon involving mean annual temperature oscillations through a range of as much as 1.5°C. It is also clear that the
cooling and warming trends were made up of shorter term oscillations which
continue into the present. Dating is not sufficiently accurate to
determine whether all the minor oscillations have been synchronous but it
is significant that the minor readvances of the 1890's, 1920's and 1960's,
about which we have detailed knowledge, were generally synchronous. It is
also possible to demonstrate that the extent of glacier advance has been
comparable globally. Glaciers in New Zealand and Patagonia and even on
Heard Island in the sub-Antarctic waxed and waned on much the same scale
as those in the Alps and Scandinavia.
        The Little Ice Age was not unprecedented. Sets of moraines comparable
in size to those formed in the Little Ice Age have been dated by radiocarbon analysis of organic material immediately below, above or within them.
In the Alps, some 7 or 8 events of similar magnitude are known to have
occurred within the last 10,000 years (9). The moraine chronology here is
supported by independent tree growth studies. Evidence of a similar type
is rapidly accumulating from other continents and the more detailed the
investigations the more numerous are the oscillations identified (10). It
is quite possible that some if not all of these oscillations were global in
extent with some regional departures from the overall pattern (11). Some
of these events earlier in the Holocene may have lasted longer than the
Little Ice Age of recent centuries.
        The fluctuations of the Little Ice Age and the similar events earlier
in the Holocene have involved small scale changes in global mean temperature which have yet to be satisfactorily explained (12). It has been shown
that volcanic eruptions can eject sufficient sulphate aerosols into the
stratosphere to cause cooling by back-scattering of solar radiation.
Analyses of annual layers from a mid-Greenland ice core back to 553 AD
provide a continuous year by year record of volcanic activity north of 20°N
(13,14). A continuous acidity profile along the ice core was measured

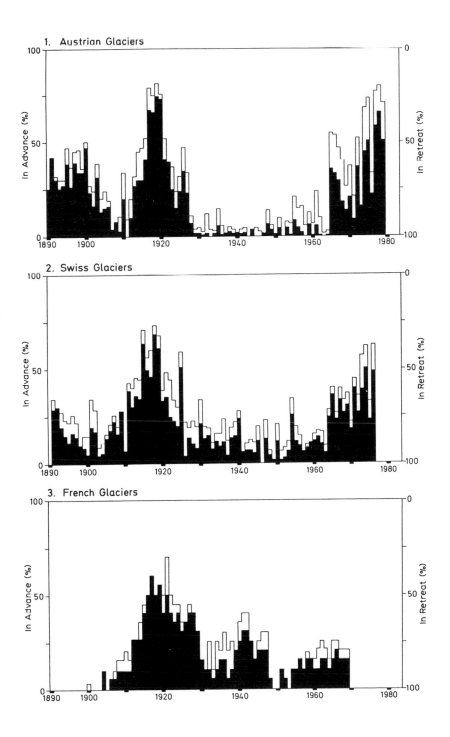

FIGURE 1 : Alpine glaciers : ● percentage advancing □percentage stationary

and compared with a volcanic activity index prepared from historic records and a northern hemisphere temperature index based on a temperature series from central England, tree-rings in the White Mountains and an isotopic series from Greenland. A significant correlation was found and the high acidity levels resulting from major eruptions such as Katmai in 1912 and Tambora in 1815 were clearly identifiable. Comparison of the glacier variations since 1870 and the acidity profiles has been found to show agreement. Furthermore, a comparison of the timing of major explosive eruptions with the available temperature data shows that such eruptions have commonly produced a small but consistent temperature decrease of the order of $0.2°$ to $0.5°C$ on a hemisphere scale for 1 to 5 years (15). However, temperatures do not remain depressed for a longer period after a series of closely timed eruptions such as those of 1881-9 and 1902-3 than they do after a single eruption. Moreover there was no decrease in temperature after some eruptions studied, and in some cases there was an increase, for example after the eruption of Bezymianny in 1956. Volcanism probably does not provide the whole answer, though it does seem to be an important factor (16). Both Rampino et al (17) and Kuhn and Shepherd (15) have found evidence of cooling preceding major eruptions by a few months!

Volcanism is not the only internal element of the global system that has been invoked as a control of climate on the time-scale we are considering. For instance both pressure distribution and sea surface temperatures during the Little Ice Age differed from the twentieth century average and changes in atmosphere-ocean relationships involving the persistence of low index circulation patterns in mid-latitudes and changes in salinity patterns in the oceans have to be taken into account (18, 19) although they do not appear to provide the essential trigger for a change of mode.

It has been suggested that variations in the earth's magnetic field and its shielding effect against solar corpuscular radiation could be involved (20, 21). Temperature changes could also be caused by changes in radiation receipt. It has been widely held that the sun is a constant star displaying regularly repeating behaviour, its character being displayed particularly clearly by the periodicity of sunspots passing across the face, but careful analysis of historic data suggests that within the last millennium the sun has been both considerably less active and probably more active than during the last 250 years (22, 23). Between 1654 and 1714, during the so-called Maunder Minimum there were very few sunspots. Changes in solar activity cause changes in cosmic ray activity which in turn affect 14C production. The resulting changes in 14C levels are recorded in tree-rings. In historic time there seems to have been a correlation between periods of low solar activity, high 14C production and low global temperature (24, 25, 26) but we have to notice that 14C levels could be modulated not only by changes in geomagnetic field intensity and solar variability but also by the exchange rate of CO_2 and 14C between the atmosphere, ocean and biosphere, about which knowledge is incomplete (23, 27).

The implications of small-scale climatic fluctuations are multifarious. The effects have been most noticeable and in some cases dramatic in high latitudes, where temperatures have oscillated most, and in the mountains. In the Alps, in Norway and in Iceland, glaciers overran farmland in the seventeenth and eighteenth centuries (28, 29). These effects were small scale in terms of the area and the numbers of people affected by the ice, but prosperity declined over much wider regions. In western Norway, the glacial advances of the seventeenth and eighteenth centuries were accompanied by a massive increase in the frequency of large landslides and major avalanches. They were on a sufficient scale to warrant tax decreases even in a time of financial stringency (30). Moreover the hay crop was so diminished that fewer cattle could be kept through the winter in this pastoral region, and the grain often failed to ripen in the "green years" of harvest failure (31).

The Little Ice Age centuries were not characterised by permanent low temperatures so much as by groups of years unfavourable for agriculture (32). Outstanding amongst them were the severe years of the 1690's when famine affected France, northern Italy, Scotland, Norway, Finland and Esthonia; even Spain and Portugal "had no such cold in many years" (33). Finland lost 30% of its population and Esthonia 20% in consequence of the severe weather and famine conditions in the years after 1695. In Norway the harvest that year failed almost completely (34).

It was especially in the oceanic northwest of Europe that the effects of sequences of cold years were felt. Here, a small change in summer warmth can have a very marked effect on the viability of cropping in upland areas. Parry (35,36) has shown that in southeast Scotland at an altitude of 300m a fall in mean summer temperature of less than 1^{c}C can result in an increase in harvest failure from about one year in twenty to about one year in three. Nearly 3000 hectares of cultivated farmland in the Lammermuir Hills reverted to moorland between 1600 and 1800. Since the late nineteenth century the rise in mean summer temperatures has brought large areas back into the zone where minimum heat requirements for the ripening of grain are met.

The vulnerability of Icelandic agriculture was forcefully demonstrated during the cold period of the 1960s and 1970s. In the period 1950-1968 a drop of mean annual temperature by 1°C involved a reduction in hay yields of one ton per hectare which in turn affected livestock production. In the single year 1967, hay production from an area of about 100,000 hectares decreased by 87,000 tons, reducing the basic productivity of the country by a fifth. This figure did not include the loss of grass on the grazing land or the resultant shortage of raw material for other agricultural production (37). The comparatively slight fluctuations of recent years, which caused only minor swelling of the Icelandic glaciers, brought the sea-ice back to Iceland, diminished the fish catch in home waters drastically, and resulted in devaluation of the currency. It is worth noticing that the greatly improved farming methods introduced in the warm years of the twentieth century have not overcome Iceland agriculture's vulnerability to climatic change.

A few years with lower temperatures have additional effects of interest to more industrialised countries. Apart from the increased costs of heating and transport, runoff from glaciers and permanent snowfields is reduced. Such runoff is crucial for the development of hydro-electric power (38). When glaciers advance, hydro-electric installations may be put at risk, as in the case of portions of the Grand Dixence scheme in Switzerland in 1983.

The events of the last two decades have shown that the "greenhouse effect" caused by the burning of fossil fuel is not so dominant as to prevent cooling over several years. Present temperatures, though higher than for several centuries, are not as high as those of early medieval times (2). The current interglacial has lasted about as long as earlier interglacials and the outcome of the conflict between anthropogenic warming and longterm cooling as it will develop over the next few decades is uncertain.

Whereas temperature is the variable of most consequence for plant growth and economic activity in highland and northern Europe, in the Mediterranean lands precipitation is of greater consequence. Its interannual variability is high and (unless large lakes are present) greater difficulties are experienced in detecting longterm trends than is the case with temperature. Furthermore, regional diversity of departure from mean conditions is greater. However there appear to be indications that, over periods of centuries, changes in temperature, distinguishable from glacier fluctuations, are associated with changes in rainfall modes as they affect river behaviour in the Mediterranean basin (39, 40). Both temperature fluctuations in the highlands and rainfall fluctuations around the Mediterranean are probably influenced by the same meteorological patterns. Such an association deserves further investigation employing historical and geological methods of research.

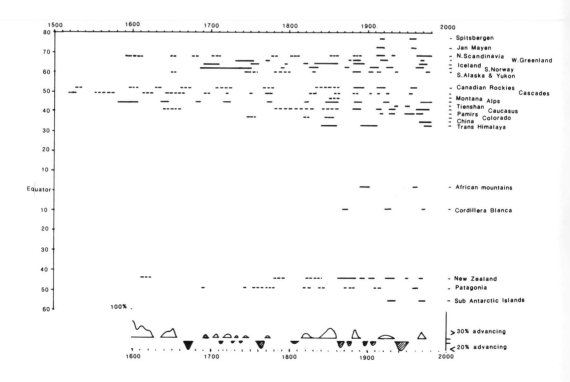

FIGURE 2 : Recorded periods of advancing glaciers

REFERENCES

1. MANLEY, G. (1974). Central England temperatures: monthly means,
    1659-1973. Quarterly Journal of the Royal Meteorological Society,
    100, 339-405.
2. LAMB, H.H. (1977). Climate, Present, Past and Future, Volume 2,
    Climatic History and the Future. Methuen, London, 835 pp.
3. VENETZ, M. (1968). Memoire sur les variations de la temperature dans
    les Alpes de la Suisse. Zurich Orelli Füssli, 33 pp.
4. REYNAUD, L.R. (1983). Recent fluctuations of alpine glaciers and
    their meteorological causes: 1880-1980. In Street-Perrott, A.,
    Beran, M. and Ratcliffe, R. (Eds.), Variations in the Global Water
    Budget, Reidel, Dordrecht, 197-205.
5. SCHWEINGRUBER, G.H., Bräker, O.U. and Schär, E. (1979). Dendro-
    climatic studies in conifers from central Europe and Great Britain,
    Boreas 8, 427-52.
6. HUGHES,M.K., SCHWEINGRUBER, F.H., CARTWRIGHT, D., AND KELLY, P.M.
    (1984). July-August temperatures of Edinburgh between 1721 and 1975
    from tree-ring density and width data. Nature, 308, 341-3.
7. PORTER, S.C. (1981). Lichometric studies in the Cascade Range of
    Washington: establishment of Rhizocarpon geographicum curve. Arctic
    and Alpine Research, 13, 11-23.
8. MILLER, C.D. (1973). Late Quaternary glacial and climatic history of
    Northern Cumberland Peninsula, Baffin Island, N.W.T., Canada.
    Quaternary Research 3, 561-83.
9. RÖTHLISBERGER, F., HAAS, P., HOLZHAUSER, H., KELLER, W., BIRCHER, W.,
    AND RENNER, F. (1980). Holocene climatic fluctuations - radiocarbon
    dating of fossil soils (fAh) and woods from moraines and glaciers in
    the Alps. In Müller, F., Bridel, L., and Schwabe, E. (Eds.),
    Geography in Switzerland, Geographica Helvetica, 35, 21-52.
10 GROVE, J.M. (1979). The glacial history of the Holocene. Progress in
    Physical Geography, 3, 1-54.
11 WILLIAMS, L.D. AND WIGLEY, T.M.L. (1983). A comparison of evidence
    for late Holocene Summer Temperature Variations in the Northern
    Hemisphere. Quaternary Research 20, 286-307.
12 ANDREWS,J.T., DAVIS, P.T., MODE, W.N., NICHOLS, H. AND SHORT, S.K.
    (1981). Relative departures in July temperatures in Northern Canada
    for the past 6,000 years. Nature 289, 164-6.
13 HAMMER, C.U. (1980). Acidity of polar ice cores in relation to
    absolute dating, past volcanism and radio-echoes. Journal of
    Glaciology, 25, 359-72.

14 HAMMER, C.U., CLAUSEN, H.B. AND DANSGAAD, W. (1980). Greenland ice
    sheet evidence of post-glacial volcanism and its climatic impact.
    Nature 288, 230-5.
15 RAMPINO, M.R. AND SELF, S. (1982). Historic Eruptions of Tambora
    (1815), Krakatau (1883) and Agung (1963). Their Stratospheric
    Aerosols and Climatic Impact. Quaternary Research 18, 127-43.
16 SELF, S., RAMPINO, M.R. AND BARBERA, J.I. (1981). The possible
    effects of large 19th and 20th century volcanic eruptions on zonal
    and hemispherical surface temperatures. Journal of Volcanology and
    Geothermal Research, 11, 42-60.
17 KUHN, G.C. AND SHEPARD, F.P. (1983). Importance of Phreatic Volcanism
    in producing abnormal weather conditions. Shore and Beach, 19-29.
18 BJERKNES, J. (1968). Atmosphere-ocean interaction during the 'Little
    Ice Age' (Seventeenth to Nineteenth Centuries A.D.). In The Causes
    of Climate Change. Meteorological Monographs 8, 77-88.

19   WEYL, P.K. (1968). The role of the oceans in climatic change: a
     theory of the Ice Ages. In Causes of Climatic Change, Meteorological
     Monographs 8, American Meteorological Society, 37-62.
20   KING, J.W. (1974). Weather and the Earth's magnetic field. Nature
     247, 131-4.
21   WOLLIN, G., ERICSON, D.B., AND RYAN, W.B. (1971). Variations in
     magnetic intensity and climatic change. Nature 232, 549-50.
22   EDDY, J.A. (1976). The Maunder minimum. Science 192, 1189.
23   STUIVER, M. AND QUAY, P.D. (1980). Changes in atmospheric C[14]
     attributed to a variable sun. Science 156, 640-2.
24   BRAY,J.R. (1965). Forest growth and glacier chronology in north-west
     North America in relation to solar activity. Nature 205, 440-3.
25   BRAY, J.R. (1967). Variations in atmospheric carbon-14 activity
     relative to sunspot-auroral solar index. Science 156, 640-2.
26   BRAY, J.R. (1968). Glaciation and solar activity since the Fifth
     Century B.C. and solar cycle. Nature 220, 672-4.
27   LAL, D. AND REVELLE, R. (1984). Atmospheric PCO_2 changes recorded in
     lake sediments. Nature 308, 344-6
28   GROVE, J.M. (1966). The Little Ice Age in the Massif of Mont Blanc.
     Transactions and papers of the Institute of British Geographers, 40,
     129-43.
29   LADURIE, E. Le R. (1972). Times of Fast, Times of Famine. George
     Allen and Unwin, 426, pp.
30   GROVE, J.M. (1972). The incidence of landslides, avalanches and
     floods in Western Norway during the Little Ice Age. Arctic and
     Alpine Research, 4, 131-8.
31   GROVE,J.M. AND BATTAGEL, A. (1983). Tax records from Western Norway,
     as an index of Little Ice Age environmental and economic deterioration.
     Climatic Change 5, 265-82.
32   PFISTER, C. (1980). The climate of Switzerland in the last 450 years.
     In F. Müller, L. Bridel and E. Schabe (Eds.). Geography in Switzer-
     land. Geographica Helvetica, 35, 21-52.
33   LINDGREN, S. AND NEUMANN, J. (1981). The cold, wet year 1695 - a
     contemporary German account. Climatic Change 3, 173-87.
34   DRYVIK, S., MYKLAND, K. AND OLDENVOLL, J. (1976). The demographic
     crisis in Norway in the 17th and 18th Centuries. Some data and
     interpolations. Universitetsforlaget, Bergen-Oslo-Tromsø.
35   PARRY, M.L. (1978). Climatic change, agriculture and settlement.
     Dawson, Folkestone, 214 pp.
36   PARRY, M.L. (1981). Climatic change and the agricultural frontier: a
     research strategy. In T.M.L. Wigley, M.L. Ingram and G. Fraser (Eds.)

     Climate and History, C.U.P., 319-36.
37   FREDRIKSSON, S. (1969). The effect of Sea Ice on Flora, Fauna and
     Agriculture. Jökull 19, 146-57.
38   COLLINS, D.N. (in press). Water and Mass Balance measurements in
     glacierised drainage basins. Geografiske Annaler.
39   VITA-FINZI, C. (1969). Mediterranean Valleys. Geological change in
     Historic Time, C.U.P., 140 pp.
40   BUTZER, K.W., MIRALLES, I. AND MATEU,J.F. (1983). Urban Geo-
     archaeology in Medieval Alzira. Journal of Archaeological Science,
     10, 333-49.

B. FACING DESERTIFICATION :

CRITERIA, METHODS AND PROGRAMMES

Productivity, variability and sustainability as criteria of desertification

A new integrative methodology for desertification studies based on magnetic and short-lived radioisotope measurements

An eco-geomorphological approach to the soil degradation and erosion problem

Hazard mapping as a tool for landslide prevention in Mediterranean areas

The "Lucdeme" program in the Southeast of Spain to combat desertification in the Mediterranean region

Soil conservation problems in Italy after the Council of Research finalized project

PRODUCTIVITY, VARIABILITY AND SUSTAINABILITY AS CRITERIA OF DESERTIFICATION

A. WARREN
Department of Geography,
University College London.

Summary

Criteria are needed to distinguish the limits of a programme of study
or control of desertification (or land degradation). These should be
as unambiguous and easy to measure as possible, but not at the expense
of their relevance to specific economies and cultures. They should
also distinguish degradation involving the environment from that which
does not. And they should specify their scale of concern: concern
with extensive and long-continued change is termed "conservation" and
is distinct from agricultural programmes. Such a programme is
properly the province of a body like the EEC. With these requirements
in mind, three sets of criteria are explored: (1) Reductions on
production or productivity; (2) increases in variability of
A production (instability), B income (inequity), or C spatial
distribution of production or productivity; and finally (3) decreases
in sustainability - the property of the system to recover from a
severe shock.

1. INTRODUCTION

A programme of action or research into either desertification or land
degradation would have to identify criteria to decide on the extent and
nature of its concern. Identification is not a simple matter and needs
careful research and debate, as we discovered in the preparation for the
UN Desertification Conference (e.g. 33). What follows is a preliminary
discussion of some promising standards for judging land degradation
whether by desertification or by some other means.

2. THE CRITERIA FOR CRITERIA

There are four sets of requirements for the criteria of environmental
degradation: clarity, relevance, environmental specificity and scale-
specificity.

2.1 Clarity

Criteria are required first to be unambiguous, objective, widely
applicable and determinable from easily gathered information. These
requirements need no further emphasis, although the discussion of relevance
and specificity below will show that wide applicability at least and
probably unambiguousness as well are unlikely to be absolutely attainable.
Because of their clarity environmental measures (albedo, groundcover,
biomass etc.) have often been proposed as criteria of desertification (10).
But their very simplicity does bring difficulties.

2.2 Relevance

Desirable as objectivity is, unambiguous environmental criteria may
not, by their very absence of subjectivity, meet a second and more important

requirement. This is that criteria should be socially relevant.

This point requires illustration. The illustration below is not European, but there are justifications for using it here. The question of relevance is best seen when two different cultures are contrasted. Although Europe is rich in different cultures most of them grade into each other. The present example displays an unusually startling contrast of ways of using semi-arid areas. It is also a useful illustration of the use of the criteria to be discussed later.

Perhaps the most apparently obvious case of desertification occurs on the Sinai-Negev border as shown in Figure 1. Otterman and his co-workers in several papers (17, 19, 20) drew attention to the contrast in albedo between the dark Negev in Israel and the light Sinai. They assumed that it reflected overgrazing i.e. desertification, in Sinai.

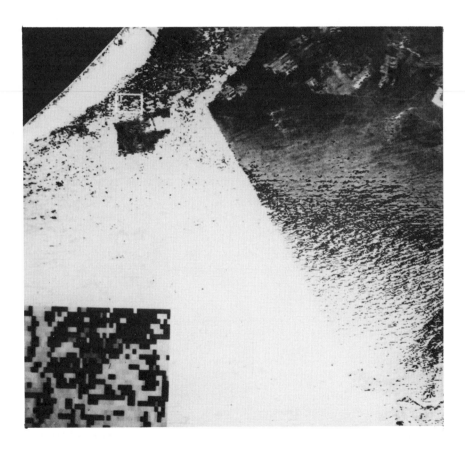

Figure 1. A LANDSAT image of the Sinai-Negev border. The internat- ional border between Israel (top right) and Egypt clearly picks out two very diffierent forms of land use. The question is: which side is degraded?, the answer is probably - both. The sustainability of the grazing system in Sinai is illustrated by the rapid re-growth of vegetation in the fenced area near the sharp bend in the border (the dark rectangular patch). This was a short-lived Israeli settlement.

There are, however, several reasons for caution in accepting such a judgement. Some of the detailed background to the argument that follows appears elsewhere (29,32); only the details relevant to the present argument are included here.

The first reason for caution is based on the ecological theory of grazing. The most comprehensive approach to this theory for areas with a restricted rainy season is by Noy-Meir (13, 14). His models, at their simplest, show a continuous decline in the yields of pasture as grazing intensity increases, with sharper declines beyond certain levels. Nevertheless, Noy-Meir (14) noted some important limitations to this simple conclusion, especially as a result of the "grazability" relationships of pastures. All the limitations suggested that it is in the interest of the grazier to keep pasture plants below their maximum biomass.

Empirical confirmation of this prediction comes from observations on an experimental farm in the Negev. As the number of sheep per hectare was increased, several measures of yield actually rose at first. Beyond an optimum stocking density most yield measures then fell again (26).

Theory and observation therefore imply that in the virtually ungrazed (darker) Negev, the taller plants may provide sub-optimal pasture productivity (see below for definition). The lighter Sinai, it is true, is almost certainly also producing pasture at a sub-optimal rate. The appearance of active sand dunes on the crests of sand ridges in Sinai and their absence in the Negev is a clear indication of this. But, if the prime interest is in pasture for stock, then the contrast in albedo across the border may not, on this argument, be between a desertified and an undesertified area, but between two areas producing sub-optimally for different reasons.

It could further be argued that mere numbers of sheep may be more important to the beduin inhabitants of Sinai than the optimal level of productivity of meat, milk or wool. Social surveys have shown that income from livestock is only a small part of the total income of beduin families, and has probably been subordinate to other income since Biblical times (3, 11, 22). But, while unimportant economically, sheep are very important social tokens (1, 2, 28). The optimization of sheer numbers, even at the expense of environmental damage and suboptimal production of meat, wool etc., may therefore be the prime objective of management among the beduin.

It could be countered that a number-optimizing strategy is causing lasting damage to the ecosystem, a point that will be discussed at more length below. However, in northwest Sinai and the Negev the evidence is that the sandy and loessic soils recover a full vegetation cover remarkably quickly when fenced against grazing (as shown by the fenced Israeli settlement in Figure 1, and in reference 18). In other words no lasting damage appears to have been inflicted by millenia of grazing.

It is true that low vegetation may not meet the needs of the Israeli settlers of the Negev. They do not rely heavily on free-range domestic stock, and plans for the area envisage large nature reserves. But caution is necessary even here. Studies in the north of Israel show that the numbers of species of birds and some invertebrates actually increases with grazing pressure by domestic stock (Rankevitch and Warburg 1983). Of course, the level of grazing that produces optimal wildlife may not be that which produces optimal economic crops.

The present relevance of the Sinai/Negev case is that actual land use is vital to the definition of criteria of degradation. In Sinai it is possible that a number-optimizing strategy or even a meat- or wool-optimizing strategy could occur with low production, high albedo and perhaps without desertification. The needs of the inhabitants of the Negev may be best served by a high level of plant cover, low productivity and low albedo, and also without desertification.

It follows that albedo or biomass can only be relevant criteria if they are related to a detailed knowledge of the objectives of management. The same argument could be applied mutatis mutandis to most environmental criteria. This is not an argument against the use of environmental indices of desertification (or land degradation), but it is a plea that their cultural relevance should be explicit in every case.

2.3 Environmental Specificity

Criteria should clearly indicate that any degradation that is identified involves environmental changes. This is by no means an easy task. The decline in the population (and by inference the productivity) of the Island of Lesbos could either be attributed to the deportation during parts of its unhappy history, or to soil degradation. In such historical explanations, and indeed in many contemporary cases, there will always be a considerable element of doubt. A well-known example of a possible mistake in the interpretation of a decline in productivity is illustrated in Figure 2.

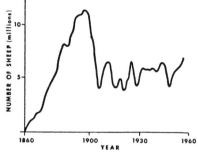

Figure 2. The trend of sheep numbers in the western Division of New South Wales. The sharp decline after about 1900 was due more to a change in market forces than to a degradation of the environment. This illustrates the care needed in discovering loss of production that specifically involves environmental degradation. (After Perry, 1968).

Declining productivity could come about in three ways. First it could be the result only of environmental change. The most easily identifiable in this category are the results of sudden disasters such as earthquakes, floods, or volcanic eruptions. Jean Grove's glacial events are of this kind (see this volume). Second it could follow a purely social, political or economic change. As with purely environmental disasters these too are hard to identify unless they are sudden: deportation, wars etc. There may be slower changes, environmental or social, that are alone the cause of declining productivity, but they usually are very hard to identify with certainty.

But much more common than either purely environmental or purely social causes of decline are those sequences of events in which social and environmental forces work together or against each other in varying mixes. A very simplified explanation of Lesbian history could be that the decline of productivity was initiated by a mass deportation or a damaging war; this would have removed the young male population that maintained agricultural terraces; neglected terraces would then encourage faster erosion. Or another simplified explanation could have it that an environmental change sapped the ability of the Lesbians to resist invasion, and that this started a similar sequence. The real explanation is certainly more complex, and involves several interactions between people and their environment. The full explanation of the curve on Figure 2 also involves such an interplay.

The pattern of change is further complicated by positive and negative feedback effects. Most usually an environmental change will invoke a negative feedback in the economic system that will tend to minimize the change: there may be new cropping strategies, or more investment of labour or capital. In such cases an identified environmental change may have no readily identifiable change in the economy. In some cases, however, there are positive feedback effects in which environmental change encourages social change and vice versa. Such was the Sahellian case in the last decade. These cases are the most obvious form of desertification or land degradation.

In many cases there may be no doubt that there is environmental involvement in a case of degradation. In many more, absolute certainty about specificity may never be attainable. With specificity, as with relevance, the requirement is more for care, and the investment of some effort in historical research, than for perfect proof.

2.4 Scale-specificity
Because changes have different meanings that depend partly on their scale, the time and space scales of the area of concern need always to be specified.

To an individual farmer a gully that appeared in one year might be a disaster. To a provincial government, only serious soil erosion on several farms over a decade would require their attention. For a national government, or the EEC, Dick Grove's proposal (this volume) that the relevant temporal scale is a generation or two, seems about right. The spatial scale for these bodies might be of the order of provinces, such as Basillicata or the Highlands of Scotland. Ideas of scale-specificity are further developed below.

Several authorities have pointed to such differences in the scale of concern of different kinds of environmental managers. The small scale is really the province of agronomists, and engineers (although these professions often show more far-sighted concern). The longer and larger view has been characterised as "Conservation". It has been argued that conservation can only be effective in the hands of bodies such as national and international governments and NGOs. Summaries of these positions can be found in references 30, 31 and 33.

3. THREE POSSIBLE CRITERIA OF DESERTIFICATION (OR LAND DEGRADATION)

In a recent paper Gordon Conway (8) proposed four criteria for the judgement of agricultural research and development programmes: production; stability; sustainability; and equity. I propose to explore the use of these as criteria in the present context, applying to them the criteria for criteria above. Stability and Equity will be combined and extended under the heading "Variability".

3.1 Production and Productivity
Productivity or production are defined here as mean values over a specified period as shown on Figure 3. A decline in mean productivity

would be the most obvious sign of desertification. It would not, however, be a sufficient sign on its own, as will be discussed below: long-term mean productivity would remain stable even if desertification were occurring.

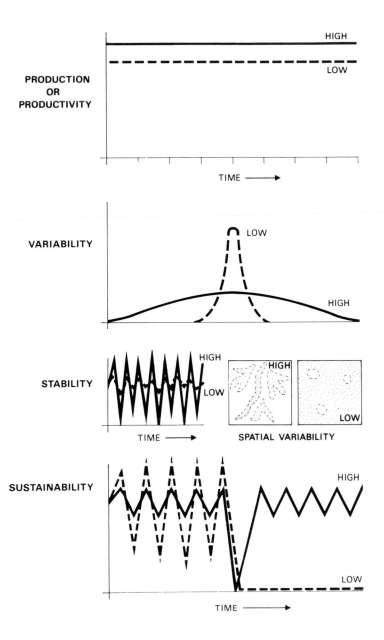

Figure 3. A diagrammatic representation of three criteria of desertification of land degradation.

3.1.1 Clarity

The most obvious need is that the product that is in decline be identified. If it is a material product (such as sheep in Sinai) this may be easy, but if it is some cultural attribute (such as amenity in the Negev) even the identification may be controversial. Data may come from official statistics, from questionnaire data, from sequential air photographs or satellite images, or from historical research. A threat to productivity would be more difficult to establish than an historical change.

It is important to distinguish between productivity (the rate of production) and production (standing crop). If the economic product is to be food or fibre, the interest lies in productivity. If the product is amenity, maximum production may be the objective.

In the case study above, it was certain that there was more plant production in the Negev, but there was uncertainty as to whether grazing productivity was greater in Sinai or in the Negev. Otterman (18) found that the contrast on the LANDSAT band that best measures growing tissue (Band 4) was less than on the other bands. This could be taken to show that productivity was not as contrasted across the border as production.

3.1.2 Environmental Specificity

Relations between a measured environmental attribute and a historical or threatened change in production or productivity are not always obvious. It is not at all clear, for example, whether high erosion rates in the Mediterranean basin are linked to declining productivity. Many soils are developed on deep, friable and relatively fertile lithologies, and these soils can erode quickly without much loss of productivity (Helen Rendell in this volume gives some examples). There is an even more tenuous link between the existence of badlands and productivity. Some badlands in Spain have been shown to have developed quickly to near their present extent after sudden events such as earth movements. Thereafter they have expanded very little (34). If this is the case, they are probably having little effect on productivity at the present. They are evidence of one catastrophic period of degradation that has now ended.

3.1.3 Scale Specificity

The size of the area over which decline is postulated must be specified for two reasons. Taking the example of gully erosion, it first may be that within a large land unit a system of gullies is an acceptable trade-off for increased productivity elsewhere, say by the adoption of mechanisation. Second, it may be that the erosion in the gullies is taking place in a zone of low value (perhaps with steep slopes and thin soils), and is contributing soil, nutrients or water to a potentially higher value area (perhaps with gentler slopes and more moisture). The Nabatean water- and soil-harvesting systems in the Negev are a case in point (35). Another example comes from the Roman farms in Libya described by Vita-Finzi (28). In northwest Europe heaths were denuded of sods and their nutrients to feed the "plaggen" soils of "in-bye" lands (22).

The time-span over which productivity is averaged is also important. If too short, then the measure would become confused with stability (see below). If too long then important changes may be missed. The argument about temporal scales applies as much as to recorded declines as to the threat of decline in the future.

3.2 Variability

In Gordon Conway's scheme stability and equity are separate, and yet are measureable by a similar parametre: variability about a mean (Figure 3).

Stability is a temporal measure of variability about a long-term average of productivity or production. Equity can be quantified as dispersion of income about the mean. The spatial distribution of production is yet another attribute of productive systems that can be measured by its dispersion from the mean. All three are worth exploring as criteria of degradation.

3.2.1 Stability

A graphical representation of temporal instability is shown in Figure 3. Some agronomists believe that stability in productivity is a prime target in agricultural development. Instability certainly makes life difficult. Moreover, there are important links between stability and sustainability. Unstable systems may pass critical thresholds beyond which the economy or environment cannot recover i.e. they become unsustainable. Nevertheless, the two ideas are distinct (see Figure 3).

Much the same arguments about clarity, relevance and environmental specificity apply to Stability as apply to Productivity. With the scale-specificity of variability, even more care is needed. First, because the level of instability is often an artefact of or is masked by the way in which official data is collected. And second, because measures of instability fluctuate wildly at smaller scales, but are almost eliminated at very large scales. It may be that the stability criterion should only be used at the meso-scale (somewhere between provinces and countries).

Semi-arid areas are often thought to be particularly susceptible to instability, largely because a level of fluctuation that would be endurable elsewhere can take semi-arid economic systems below a sustainable threshold. Instability is therefore a specially useful criterion for desertification.

Economic systems, especially in semi-arid areas, assign a very high priority to combatting instability with strategies such as nomadism and storage. For this reason there can be strong positive feedback processes when things begin to go quite unexpectedly wrong. A delicately balanced edifice can crumble quickly. Economic instability can feed environmental instability and vice versa. Instability, therefore, whether economic or environmental, is always a useful indicator of desertification, if only potential.

3.2.2 Equity

In an environment in which it is relatively easy to make a living it should be possible to maintain relative equity (though many hierarchial social systems prevent this). But in an already harsh environment such as the desert edge, where moreover instability may often reduce the economy below a threshold of viability, inequity may be unavoidable. The evidence of nomadic economies supports this model. Although there is considerable contemporary debate about egality in nomadic societies (Bonte 1981), there is no doubt that in each harsh year a large proportion of nomadic families find that they cannot provide for themselves, and having failed, are either supported by the rest of the group or leave the nomadic group altogether (4, 7, 25).

The environment, however, is far from being the only control of inequity, so that equity can only be a criterion of degradation if all other controls are seen to be equal. On the other hand, if environmental change can be shown to be associated with increasing inequity then this might be seen as one of the strongest arguments for an action programme that involved environmental modification.

3.2.3 Spatial Variability

One of the most salient characteristics of semi-arid ecosystems is that their production and productivity are highly spatially variable (e.g. 6, 9).

A diagramatic representation of spatial variability is shown in Figure 3.
The chief reason is that the availability of water is very spatially variable by reason of topography. It may be that increasing spatial variability,
if provable, say on satellite imagery, might serve as an index of degradation.
The criterion may also hold for any system in which a critical threshold is
near and controls production and productivity: for example for forests
damaged by acid rains, uplands degraded by nutrient loss etc.

3.3 Sustainability
    Sustainability is the ability of a natural system to recover from a
severe shock (Figure 3). It is only a viable concept in the long-term, and,
as shown below, usually only over a large area. It is certainly a criterion
to be considered by an authority concerned with large rather than small
times and areas. Almost by definition, the shock in question would be
inflicted by some form of misuse or exploitation by people. Natural systems
are either assumed to be in some kind of long-term equilibrium with a
variable environment or to change irreversibly with extreme changes in
external inputs.
    Most authorities assume, many tacitly, that sustainability is the most
important criterion of degradation. The main plank of the environmentalist
argument in the 1970s has been that sustainability is being damaged by
contemporary environmental management (15). But, while vital concerns,
sustainability or its absence are neither simple criteria, nor easy to
establish (16).
    There is a theoretical distinction between the ability of the environ-
ment to sustain a specified form of land use ("environmental sustainability")
and the sustainability of the economy that depends on that environment
("economic sustainability"). The Relevance of environmental sustainability
to the particular form of land use, and its distinction from economic
sustainability must always be made in the interests of Clarity. But though
theoretically distinct, the two are in practice hard to disentangle because
of the dynamic nature of land-use systems, not least under the influence of
environmental change. Proving environmental sustainability may be difficult,
but it is easy compared to establishing the sustainability of that economic
system. Moreover, there are strong feedback processes between the two
notions. These points need illustration.
    If the land-use system in the northwestern Negev and Sinai is
defined as nomadic grazing (as practised for millennia), it seems
plain from the evidence cited above, that pastures subjected to such
treatment can quickly recover their vegetation. It is not clear
whether the recovered vegetation would contain the same species-mix as
it did formerly, and to some users this may be a more important
attribute than mere cover. But these characteristics can be determined.
    Some environments, like the sandy soils in the northwestern Negev
and Sinai, are inherently much more "sustainable" than others. On
the hard thin soils of the Negev Highlands pastures enclosed against
grazing show no recovery even after several years. The sands, unlike
the hard soils, absorb water, rather than yielding it to run-off, and
they retain seed in their friable surface layers. Sandy soils in semi-
arid areas world-wide seem to have these "sustainable" characteristics.
    When attention moves from their pastures to the whole beduin
economy, the problems of establishing sustainability are vastly multi-
plied. It could be claimed that the beduin economy has proved its
sustainability by surviving for over 6000 years in the same area. The
truth of this claim depends on historical demographic data that are
absent. But even if sustainable, the beduin economy depended, before

1949, on sources of income from beyond the semi-arid area (such as fees from caravaneering, protection and guidance of travellers, bad-season grazing in nearby wetter areas, smuggling, wage-labour, booty from raids etc.). In other words, the economy depended for its sustainability on a large area.

The mix of these and local environmentally-based sources of income was always changing, and each change raised new questions about environmental sustainability. After 1949 the loss of access to areas across the border removed some of these external sources of income, and it could be that the whole beduin economy is suffering as a consequence. This stress may be reflected in the state of the pastures, some of which are apparently being grazed more intensively because of the loss of seasonal grazings over the border, and because of the need to generate more income from this sector of the economy. Recent studies of the recovery of the sandy pastures suggest that even this system of grazing may be sustainable (18), although it is uncertain for how long.

The extreme difficulty of defining the sustainability of an economy, and the relation of economic to environmental sustainability is even clearer with the Israeli agricultural economy in the northwestern Negev. The sustainability of this energy-intensive, highly commercial economy depends on prices of energy, the maintenance of external markets, international relations etc. The time- and space-scales of the economy on which the Israelis depend for sustainability, therefore, are even larger than that of the nomadic beduin. In turn, the environmental sustainability of the resource-base of the Israeli settlers depends on the level and kinds of input from capital, labour and technology from this extended economy, and the mix of these is always in flux.

The difficulties with the notion of sustainability mean that it, like the other criteria, can probably never be proven or disproven. Any investigation of sustainability would have to involve continual iterations between environmental and economic sustainability, and results could never be taken to be permanently established.

4. CONCLUSION

Productivity, variability and sustainability offer a promising start for assessing the need for research programmes into actual cases of reputed desertification or environmental degradation. None can be established with high levels of proof, but they should at least be a check-list of questions to be asked at an early stage in a programme, and at best should themselves be urgent cases for research funding.

REFERENCES

1. 'AREF EL 'AREF (1933). Justice among the beduins. Beit ul Makden Press, Jerusalem.
2. 'AREF EL 'AREF (1944). Beduin love, law and legend, dealing exclusively with the Badu of Beersheba. Cosmos Publishing Co. Jerusalem, 207 pp.
3. BAILEY, C. (1980). The Negev in the nineteenth century - reconstructing history from beduin oral history. Asian and African Studies, 14, 35-80.
4. BIRKS, T.S. (1981). The impact of economic development on pastoral nomadism in the Middle East: an inevitable eclipse? in Change and Development in the Middle East ed. Clarke, J.I. and Bowen-Jones, H. Methuen, London, 82-94.

5. BONTE, P. (1981). Les eleveurs de l'Afrique de l'Est, sont-ils egalitaires? A propos traveaux recents. Production Pastorale et Societe, 8, 23-37.
6. BOURLIERE, F. and HADLEY, M. (1970). The ecology of tropical savannas. Annual Review of Ecology and Systematics, 1, 125-152.
7. BRIANT, P. (1981). Territoires nomades au Moyen-Orient ancien: les representations antiques. Production Pastorale et Societe, 8, 50-52.
8. CONWAY, G. (1984). Rural resource conflicts in the U.K. and the Third World. Science Policy Research Unit, University of Sussex.
9. FLORET, C. and LE FLOC'H, E. (1973). Production, sensibilite et evolution de la vegetation et du milieu en Tunisia presaharienne. CNRS, CEPE, Montpellier, No.71, 45 pp.
10. MACLEOD, N.H., SCHUBERT, J.S. and ANAEIJONU, P. (1977). Report on the Skylab 4 African Drought and arid lands experiment. in Skylab Explores the Earth, NASA SP-380, 263-286.
11. MARX, E. (1967). Beduin of the Negev. Manchester University Press, 260 pp.
12. MARX, E. (1978). The ecology and politics of nomadic pastoralists in the Middle East. in The Nomadic Alternative ed. Weissleder, W., Mouton, The Hague, 41-74.
13. NOV-MEIR, I. (1975). Stability of grazing systems, an application of predator-prey graphs. Journal of Ecology. 63, 459-481.
14. NOY-MEIR, I. (1978). Grazing and production in seasonal pasture: analysis of a simple model. Journal of Applied Ecology, 15, 809-835.
15. O'RIORDAN, T. (1983). Putting trust in the countryside. in The Conservation and Development Strategy for the U.K. Kogan Page, London, 171-260.
17. OTTERMAN, J. (1974). Baring of high albedo soils by overgrazing: a hypothesised desertification mechanism. Science, 86, 531-533.
18. OTTERMAN, J. (1977a). Monitoring surface albedo change with LANDSAT. Geophysics Research Letters, 4, 441-444.
19. OTTERMAN, J. (1977b). Anthropogenic impact on the albedo of the earth. Climatic Change, 1, 137-155.
20. OTTERMAN, J., OHRING, G. and GINZBURG, A. (1974). Results of the Israeli multidisciplinary data analysis of ERTS-1 imagery. Remote Sensing of Environment, 3, 133-148.
21. OTTERMAN, J. WEISEL, Y. and ROSENBERG, E. (1975). Western Negev and Sinai ecosystem: comparative study of vegetation, albedo and temperatures. Agro-Ecosystems, 2, 47-59.
22. PAPE, J.C. (1970). Plaggen soils of the Netherlands. Geoderma, 4, 299-255.
23. PERRY, R.A. (1968). Australia's arid rangelands. Annals of the Arid Zone, 7, 243-247.
24. RANKEVICH, Dina and WARBURG, M.R. (1983). Diversity of bird species in mesic and xeric habitats with the Mediterranean region of northern Israel. Journal of Arid Environments, 6, 161-171.
25. ROWTON, M.B. (1973). Autonomy and nomadism in western Asia. Orietalia, 42, 247-258.
26. TADMOR, N.H., EYAL, E. and BENJAMIN, R.W. (1974). Plant and sheep production on semi-arid annual grassland in Israel. Journal of Range Management, 27, 427-432.
27. TAHAL CONSULTANTS LTD. (1974). Centralized Services for the Beduin in Sinai.(In Hebrew) Israeli Military Government, Sinai Region H.Q.
28. VITA-FINZI, C. (1969). The Mediterranean Valleys: geological changes in historical time. Cambridge University Press, Cambridge, 140 pp.

29. WARREN, A. (in prep.). Science and culture in the management of dry lands. Chapter 2, Nomads, Case Study - The Negev and Sinai.
30. WARREN, A. and GOLDSMITH, F.B. (1973). An introduction to conservation in the natural environment. in Conservation in Practice ed. Warren, A. and Goldsmith, F.B., John Wiley and Sons, Chichester, 1-12.
31. WARREN, A. and GOLDSMITH, F.B. (1983). An introduction to nature conservation. in Conservation in Perspective ed. Warren, A. and Goldsmith, F.B., John Wiley and Sons, Chichester, 1-15.
32. WARREN, A. and HARRISON, Carolyn M. (1984). People and the ecosystem, biogeography as a study of ecology and culture. Geoderma.
33. WARREN, A. and MAIZELS, Judith K. (1977). Ecological change and desertification. Background Paper A. Conf./74/7. U.N. Conference on Desertification, U.N.E.P. Nairobi. Also in Desertification, its causes and consequences, ed. U.N. Desertification Secretariat, Pergamon, Oxford, 169-260.
34. WISE, S.M., THORNES, J.B. and GILMAN, A. (1982). How old are the badlands? A case study from south-east Spain. in Badland Geomorphology and Piping, ed. Bryan, R.B. and Yair, A., Geobooks, Norwich, 259-277.
35. YAIR, A. (1983). Hillslope hydrology, water harvesting and arial distribution of some ancient agricultural systems in the northern Negev desert. Journal of Arid Environments, 6, 283-301.

A NEW INTEGRATIVE METHODOLOGY FOR DESERTIFICATION STUDIES BASED ON MAGNETIC AND SHORT-LIVED RADIOISOTOPE MEASUREMENTS.

F. Oldfield, S. R. Higgitt, B.A. Maher,
Dept of Geography
P.G. Appleby,
Dept of Applied Maths
University of Liverpool, P O Box 147, Liverpool L69 3BX, U.K.

M. Scoullos,
Dept of Chemistry, University of Athens, Greece.

SUMMARY

The watershed-ecosystem concept provides a spatially bounded framework within which many aspects of material flux and consequent ecological change can be characterized and quantified, and their interactions studied on a wide range of temporal and spatial scales. Thus in any analysis of the environmental aspects of desertification, lakes and near-shore marine environments provide some of the most favourable opportunities available. This is especially so with the advent of several new techniques using magnetic and short lived radioisotope measurements. The present account focuses on these new techniques and their applications. 'Mineral' magnetic measurements can be used to (i) speed and enhance quantitiative estimates of both past and present sediment yields in eroding catchments (ii) identify past and present sediment sources (iii) establish areas of soil depletion and redeposition and characterize associated slope processes (iv) provide a basis for active tracing experiments on slopes and in river channels, lakes and coastal environments and (iv) characterize atmospheric dusts and ascribe them to source type. Studies using the short lived natural radio-isotope lead-210 (half life 22.26/yr) can provide chronologies of sedimentation for the last 100 to 150 years. Used in conjunction, these methods, alongside more conventional geomorphological, sedimentological, palaeoecological and geochemical techniques, can form the core of an integrated multidisciplinary study of desertification and erosion processes on all relevant temporal and spatial scales.

1.0 THE RATIONALE FOR SEDIMENT BASED STUDIES OF DESERTIFICATION

Understanding, monitoring and ameliorating ecological degradation whether caused by the natural course of environmental change, the excesses of human exploitation, or some conjunction of the two, implies measuring the dynamics of change through time especially with regard to limiting or forcing variables within the ecosystem in question. Until recently, most studies of ecological changes dealt either with contemporary observations and experiments or with historical reconstruction on much longer time scales during the past. Relatively few studies concentrated on linking insights gained from short term observations and experiments with those obtained from the reconstruction of past conditions and variations on longer timescales. In recent years and in recognition of this crucial gap, an increasing number of workers have begun to explore contexts within which a continuum of insight into environmental change can be gained.

For a true 'continuum of insight' we are often dependent on situations where more or less continuous accumulation has taken place, whether this be of ice, peat or sediment. In each type of context a growing range of new techniques has made possible linking studies in which the present is not seen merely as a key to the past, nor is the past seen as a rather poorly defined historical context for the conditions of the present. **Instead the detailed reconstruction of a sequence of changes in the recent past is used in the direct interpretation and evaluation of contemporary problems, These problems and their future development can be seen as the expressions of processes and trends which can be dated, interpreted and often quantified from historical evidence.**

Lakes and near shore marine environments, and the sediments which accumulate within them comprise a uniquely valuable source of evidence for reconstructing the course and pattern of degradation, as well as its causes and effects. The basic rationale underlying the study of recent lake and marine sediments is set out briefly in Oldfield (1977 and 1983b). Figure 1 summarises the linkage between the empirical field studies implied and the problems of modelling and ultimately decision making involved in desertification studies.) Having chosen sediments and the watersheds from which they derive as our context for study, we need techniques which assist in the establishment of chronologies of sedimentation so that events can be dated and rates of change can be calculated. We also need to develop methods of analysis using conservative and diagnostic products of environmental and ecological processes, whether these products are the chemical or magnetic characteristics of mineral material in the sediment, or the preserved remains of environmentally indicative or ecologically significant organisms. Moreover, there are overwhelming arguments for using these methods in conjunction.

2.0 THE MINERAL MAGNETIC APPROACH TO EROSION AND SEDIMENTATION STUDIES

2.1 INTRODUCTION

Mineral magnetic studies of lake, estuarine and near-shore marine sediments began around 1970 and grew out of the palaeomagnetic work of Mackereth (1971) and Thompson (1973). Their pioneer studies of recent palaeomagnetic secular variation were designed to provide proxy timescales of sedimentation first in Windermere then in Lough Neagh. Coincidental to this work came the demonstration that in the Lough Neagh sediments, magnetic susceptibility could be used to synchronise cores and, moreover, showed a strong direct correlation with variations in the relative frequency of pollen types indicative of deforestation and farming (Thompson et al. 1975). This suggested that magnetic properties could be used both for improving quantitative estimates of sedimentation and for relating changes in sedimentation to catchment processes.

Since the initial magnetic studies of Lough Neagh were carried out, many hundreds of cores from over 80 lakes and estuaries have been measured. The range of 'mineral' magnetic (as distinct from palaeomagnetic) measurements has grown from simple volume susceptibility to an extended sequence of 'in-field' and laboratory-induced remanence measurements capable of providing detailed sample characterization and differentiation. Specially designed equipment has been developed for the field and laboratory measurement of surfaces, whole cores, sub-samples of sediment whether moist or dry, and filtered suspensates. The sites studied cover a wide variety of lithologies, range from Arctic-alpine to subtropical in climatic regime, and include not only lakes of all shapes and sizes, but also estuarine and other near shore depositional environments. The thrust of the work completed so far has been towards:

1. developing a coherent mineral magnetic methodology for drainage basin erosion and sedimentation studies (2.2 below)
2. exploring the value of the mineral magnetic approach to core correlation and the quantification of sedimentation and erosion (2.3)
3. testing and refining the basis on which source-sediment linkages can be established from mineral magnetic evidence in as wide a range of climatic and lithological contexts as possible (2.4)

and

4. developing and evaluating magnetic tagging and tracing techniques (2.5)

2.2 METHODOLOGY

Oldfield (1983)a and Oldfield and Maher (1984) outline measuring sequences for catchment based magnetic studies. The measurements begin with rapid scanning techniques for catchment surfaces and for core material, pass through an ordered succession of susceptibility and remanence determinations designed to provide a basis for detailed characterization, and conclude with various types of non-magnetic and often sample destructive analysis.

Measurements of natural remanence can be included in the sequence for those situations where timescales may be derived from palaeomagnetic analysis. This applies to many lake and near shore marine environments as well as to loess and palaeosol sequences. In the case of sediment cores, initial natural remanence measurements as well as susceptibility scans can often be carried out before extrusion from the core tube and appropriate dates can be obtained very rapidly indeed. The availability of specially designed equipment together with the ease and speed of measurement makes much of this work feasible as a field excercise prior to further studies in the laboratory.

2.3 CORE CORRELATION AND SEDIMENT YIELD CALCULATIONS

The first attempt to use magnetic correlations to calculate total sediment input for defined time intervals was carried out for the recent sediments of Llyn Goddionduon, N. Wales, using a Mackereth mini corer to take well over a hundred cores which were scanned for volume susceptibility variations. The main correlating horizons were linked by not simply curve-matching but by carrying out detailed supplementary mineral magnetic measurements on key cores, and critical horizons were dated by ^{14}C and [137]Cs. A preliminary outline of the approach is given in Bloemendal et al (1979), a full account of the dated sediment influx variations on a whole-lake basis are given in Bloemendal (1982) and summarised in Thompson & Oldfield (in press). The main mineral magnetic variations used in correlation were interpreted as resulting from water level changes and a recent forest fire in the catchment.

Subsequent work by Dearing et al. (1981) at Llyn Peris, N. Wales used a similar approach based on a much lower density of cores but with a detailed chronology provided by ^{210}Pb and palaeomagnetic secular variation. In this case, the variations in magnetic susceptibility upon which core correlations were based were interpreted as reflecting shifts in dominant land use in the large catchment as a result of vicissitudes in upland grazing and slate quarrying over the last two centuries.

A more closely controlled study by Dearing (1983) at Havgardsjon, S. Sweden, using some 50 cores, confirming mineral magnetic correlation by comparison with visual stratigraphic changes and dating the sequence of changes by palaeomagnetic and radiometric methods, has provided a basis not only for sediment influx calculation but for historical nutrient budgets within the catchment (Oldfield et al. 1983). Here, as in the Lake Egari study in the New Guinea Highlands (Oldfield, et al. 1980 and in press), the main variations in sedimentation rate are interpreted as reflecting agricultural changes within the catchment. The Lake Egari study leans rather more heavily on radiometric data than on mineral magnetic correlations, but it includes some attempt to list criteria for identifying lakes suitable for sediment-based catchment erosion studies.

In the four cases noted above, some attempt has been made to use dated and correlated core sequences to provide estimates of allochthonous sediment yield from the lake catchment for defined recent time intervals, and in this sense, they illustrate attempts to use magnetic studies of lake sediments as a basis for the quantitative reconstruction of historical erosion rates and their relation to catchment processes resulting largely from human activities. The methods are ideally suited to studies of ecological degradation. There is no reason why this approach should not be applied on a much larger spatial scale to basins such as the N. E. Mediterranean itself and preliminary results (Thompson and Oldfield, in press) are very encouraging.

2.4   SEDIMENT SOURCE IDENTIFICATION

Following the work on suspended sediment source identification in the Jackmoor Brook (Oldfield et al. 1979; Walling et al. 1979), many other studies have been carried out in a variety of fluvial, lacustrine and estuarine environments. In addition, mineral magnetic differentiation in the regolith on different lithologies and under a variety of climatic conditions has been intensively studied. The most detailed magnetically based catchment study of erosion and sedimentation so far has been carried out in the Rhode River Watershed. The Rhode River is a tidal arm of Chesapeake Bay which, along with its catchment, has been intensively monitored for over a decade by the Smithsonian Institution. Preliminary results are reported in Oldfield (1983), Oldfield et al (in prep.) and Thompson & Oldfield (in press). Throughout the catchment, on a variety of sedimentary lithologies, soil profile development has given rise to changes in mineral magnetic properties which can be used to characterize each horizon. The parameters used to identify each horizon can also be recognized in the suspended sediments of the streams and the estuarine waters, as well as in the estuarine sediments themselves. They allow confident distinction between soil- and substrate-derived sediments and this distinction can be established using mutually independent magnetic parameters; moreover it can be sustained in all phases of the system on a particle-size specific basis. Where the sediment is derived from well differentiated and deeply weathered soils, the dominant magnetic component of each of its particle size fractions can often be ascribed to the particular soil horizon from which it came. Studies in progress elsewhere including the Lac d'Annecy in France (Fig. 2), as well as the Acheloos estuary and the Elefsis, Maliakos and Amvrakikos Gulfs in Greece (Oldfield and Scoullos, in press and Fig. 3), confirm that similar source identification can be established on virtually every type of lithology provided that (1) some weathering and soil formation have taken place, (2) the timescale over which erosion and sedimentation are being studied is substantially shorter than the rate at which the assemblage of metastable iron oxides is being transformed in the potentially eroding soil and (3) sediment diagenesis has not affected the magnetic properties of the eroded material. The method is thus particularly suited to recent rapidly eroding systems. It has proved especially useful in identifying the impact of fire (Rummery, 1983) deforestation (Oldfield et al 1980) and commercial afforestation (Appleby et al. in press). Optimism about the general applicability of the method derives from lake- and estuarine-sediment based studies carried out on a wide variety of sedimentary, igneous and metamorphic lithologies and from measurements carried out on soil profiles formed under most of the major world climatic types. In all but the coldest or most arid environments, weathering and soil formation, as well as fire, give rise to very fine ferrimagnetic iron oxide crystals within and just below the stable single domain size range (< 0.1 to 0.03μm). This phenomenon of magnetic enhancement in surface soils is especially well marked in Mediterranean lands. The fine ferrimagnetic oxides can be recognized by their relatively high anhysteretic remanent magnetization (ARM) and/or quadrature (or frequency dependent) susceptibility $X_q(X_{fd})$. In the arid soils studied these effects are less noticeable but the increased anti-ferromagnetic content of surface material arising from secondary haematite and/or goethite formation is often readily observed. There are therefore relatively few lithological contexts within which mineral magnetic parameters fail to distinguish 'soil' from parent material, and many in which a much more detailed vertical differentiation is feasible.

2.5   MAGNETIC TAGGING AND TRACING

Just as the naturally evolved magnetic properties of soils provide a basis for sediment source identification, artificially induced magnetic characteristics can be used to provide material for use in tracing experiments. As we have seen in Chapter 8, fire can lead to a strongly enhanced magnetic signal in surface soils. In practice, most reasonably iron-rich natural materials can be magnetically enhanced by heat treatment in the laboratory, though the initial idea for magnetic tagging and tracing came from monitoring the after effects of a major forest fire. The Llyn Bychan forest fire of 1976 (Rummery, 1981, 1983) in N. Wales gave rise to magnetically enhanced material which persisted in the soils and the lake sediments and found its way to the Afon Abrach, the river which drains both the lake and the intensively burnt area downstream of the lake outfall . After the fire, a series of magnetic measurements were made on sieved material from successive downstream shoals beginning in the burnt area and continuing for some 2 km. In the case of the coarser clasts the downstream magnetic variation involves an order of magnitude decline. By contrast, the finest materials show exceptionally high magnetic concentrations close to the fire and a three order of magnitude decline downstream. These results were interpreted as reflecting the selective loss of magnetically enhanced fines from the burnt area and the gradual dilution of this material at increasing distances downstream. This observation opened up the possibility of using not only naturally but artificially enhanced material as a bedload tracer in river channels.

The Plynlimon area of central Wales was chosen for the initial testing of magnetically enhanced tracers for several reasons. One of the major concerns of the Institute of Hydrology's catchment research at Plynlimon is the large volume of bedload generated in the upper reaches of the Severn by the rapid recent erosion of forest drainage ditches. By the time the first magnetic trials began, a major programme of hydrological and sedimentological monitoring had been established by the Institute and this provided an essential framework for the trials. More recently, the work was extended downstream into the piedmont zone in response to concern expressed by the Ministry of Agriculture Fisheries and Foods about the effects that the increased gravel yields, coupled with the water regulation policies adopted in headwater reservoirs, might be having on channel stability and possible land loss in cultivated areas.

The main reasons for using bedload tracer studies as part of the research strategy devised in response to the academic and practical problems posed by the Upper Severn are set out in Arkell et al (1984). Most conventional tagging and tracing techniques are limited to a particular particle size range. Pebble painting or plugging with radioisotopes are suitable only for large clasts, whereas fluorescence is more applicable to sands. Existing techniques also pose serious problems of signal persistence and particle recovery. Problems are further compounded in the study area by the very wide size range of the bedload and the preponderance in many reaches of fine gravel which is difficult to tag conventionally. Fortunately the bedrock is a shale uniformly rich in finely disseminated paramagnetic iron giving both a consistently low susceptibility and saturation remanence in its unheated state and a high potential for enhancement by heat treatment.

The work completed so far has involved developing
(1)   suitable heating procedures for enhancing the magnetic susceptibility of large quantities of gravel,
(2)   instruments and techniques for magnetic measurements both in the river channels and on abstracted material in the laboratory and
(3)   field trial strategies on a range of spatial and temporal scales designed to contribute both to the evaluation of the technique and to the understanding of the substantive problems of bedload transport in the area.

The heating trials leading up to adoption of a practical bulk 'toasting' method for the Plynlimon material are described and their mineralogical effects interpreted in Oldfield et al. (1981). New instrumentation was developed for the project in the form of a 20 cm diameter search loop and both hand held and ground search versions of ferrite probes constructed by Bartington Instruments. The main hydrological implications of the results obtained from all the traces so far are set out in Arkell et al (1982), and Arkell et al (1984).

Further development of the magnetic tracing methods used at Plynlimon and their adaptation to river, lake and estuarine sediment tracing on a wider range of lithologies will be constrained by several factors, some environmental and some technical. The main ones are outlined in Arkell et al. 1982 and Thompson and Oldfield, in press).

In the context of the present themes, magnetic tracing of contemporary particle movement on slopes, in river channels and in near shore environments offers new opportunities for process monitoring in degrading areas. The promise of the technique has recently been greatly expanded by initial studies confirming the suitability of several types of cheap non toxic, highly magnetic and widely available industrial waste as tracer materials across a wide range of particle sizes.

3.0   APPLICATION OF LEAD-210 MEASUREMENTS TO EROSION STUDIES

An important new development in erosion and sedimentation studies during the past few years has been the application of ^{210}Pb analyses to the dating of lake, estuarine and near shore marine sediments, and the calculation of net sedimentation rates. A complementary way in which ^{210}Pb can be used is in the identification of areas of surface soil or sediment depletion and enrichment within the catchment or lake. To date, however, this second kind of study has received little attention, (but see Section 4.0).

The main pathways by which lake sediments accumulate unsupported (or excess)^{210}Pb are illustrated in Fig. 6. It is likely that different processes will be dominant at different sites or at the same site at different times, and there will accordingly be uncertainties about the appropriateness of the various procedures for calculating ^{210}Pb dates. In spite of this the technique remains of crucial value in erosion studies since it is often the only one available which holds out any promise of establishing varying rates of sedimentation over the last 100-150 years with a reasonably fine temporal resolution. Most other techniques provide, at best, one or two dating horizons, and only where demonstrably annual laminations are identifiable is a clearly superior alternative to ^{210}Pb available. In any attempt to place contemporary effects and future impacts into longer time perspective, past rates as well as dates are essential, and in this respect, ^{210}Pb can often make a crucial contribution to establishing the continuum of insight into past and present processes which is a vital ingredient in environmental monitoring (cf. Oldfield, 1977, 1983b).

It is clear from Figure 4 that the distribution of ^{210}Pb in lake sediments may be determined by many factors, and the calculation of a reliable chronology will be feasible only if one of these factors is dominant, and readily identifiable. The principal models considered in the literature to date are the CRS (constant rate of ^{210}Pb supply) models, and the CIC (constant initial ^{210}Pb concentration) models. In the CRS models it is assumed that the dominant processes result in a constant rate of supply of ^{210}Pb to the sediments irrespective of any variations in the sedimentation rate. This is usually assumed to be associated with a situation in which the direct input from the atmosphere is dominant. The CIC models assume that in any given profile all the sediments from each depth have the same initial ^{210}Pb concentrations at the time of their formation. This may occur where the input of ^{210}Pb via eroded catchment surface soils is dominant. Variants of both models can be devised to take account of postdepositional processes such as sediment mixing or redistribution. Where dry-mass accumulation rates have been constant both models predict a simple linear relation between declining excess ^{210}Pb concentration and the cumulative dry-mass (when plotted on a log-normal scale), and give identical results. Where the sedimentation rate has varied, the dates and sedimentation rates obtained will depend on the model used. The work of the present research group rests on an approach to ^{210}Pb dating in which emphasis is placed on empirical testing at each stage in the study. From a data base which comprises ^{210}Pb records for about 100 cores from 50 lakes Oldfield and Appleby (1984a) have developed criteria for assessing whether a ^{210}Pb data set lies at one of the two poles represented by the CRS and CIC models (Appleby and Oldfield, 1983). For each core, dates and sedimentation rates are calculated using, wherever possible, alternative CRS and CIC assumptions. If the data do not conform to either of the two models the two sets of results help to set bounds on the possible chronology. Where neither model can be validated, many authors simply calculate a mean sedimentation rate from a least squares fit to the concentration versus accumulated dry mass values. In practice, this implies assuming roughly constant flux and constant sedimentation rate.

In recent reviews Oldfield and Appleby 1984 (a, b and c) have put forward (i) a series of arguments for accepting the CRS model as the primary dating model at a wide range of sites, (ii) several proposals for building empirical testing into dating strategies wherever possible, (iii) methods for building modifications into the dating model to accommodate processes such as sediment focusing and bioturbation and (iv) suggestions for areas of further research in ^{210}Pb dating. Of all the areas of further research proposed, the one of most direct relevance to the present Symposium concerns the flux of excess ^{210}Pb from catchment surfaces as part of the erosive input to the lake or estuary.

4.0 THE USE OF CAESIUM-137 IN EROSION STUDIES

Caesium-137 (half-life 30 years) is a product of nuclear explosions and so has been present globally in the atmosphere only since 1952. The first significant increase in detectable fallout in the Northern Hemisphere occurred in 1954. Peak fallout years occurred in 1959 and, especially 1963 since when, as a result of test ban treaties, fallout has declined. Its distribution can be used in two complementary ways in erosion and desertification studies. The changing caesium-137 concentration record in lake and marine sediments can often be used as a direct dating technique. Usually it provides just one or two dated horizons reflecting, for example the initial 1954 increase and/or the major 1963 peak.

Especially where sediments have accumulated slowly, diffusion of ^{137}Cs in the pore waters of the sediment can introduce important errors. Where a significant proportion of the ^{137}Cs input to sediments has been derived from eroding surface soil this will have the effect of increasing the total ^{137}Cs inventory at the site of deposition (see below) as well as blurring concentration features above the initial increase. Both effects, as well as bioturbation, can reduce the value of sedimentary ^{137}Cs as a dating tool. However, the tendency for ^{137}Cs to be redistributed within a catchment not in solution but attached to fine soil particles gives it additional value in erosion studies.

Once the total input to a given site or area can be estimated, comparison of total soil ^{137}Cs inventories with each other and with the atmospheric input allows identification of areas of net depletion and enrichment. These in turn will be   an expression of the integrated effect of erosion and redeposition processes since the mid 1950s. A general review of this approach and an elegant illustration of its application is given in Longmore et al. (1983). It is well suited for combination with both mineral magnetic and lead-210 catchment based studies of soil erosion and ecological degration.

5.0   SUMMARY AND PROJECT SUGGESTIONS

1.   Sediment based studies of desertification can, within the ecosystem watershed conceptual framework, provide not only a historical background to contemporary problems but also a continuum of insight into their immediate antecedents, an empirical input into modelling their effects and a dynamic baseline for monitoring their future impact.

2.   In this connection,
          (a)   magnetic measurements can provide
          -     timescales of sedimentation (10^2 - 10^4 years)
          -     bases for quantifying material flux within degraded ecosystems
          -     evidence for contemporary and past sediment source types,
          -     confident identification of fine soil-derived particulates
          -     identification of dust types and sources
          (b)   short-lived radioisotope studies (^{210}Pb; ^{137}Cs) can provide
          -     detailed timescales of sedimentation (1 - 10^2 years)
          -     evidence for the recent redistribution of surface materials within catchments
          -     identification of dust types and sources.

3.   Used in conjunction with geochemical and palaeoecological techniques these new methods provide a powerful new basis for understanding the dynamics of ecosystem change and especially for resolving major questions regarding the relative contribution of climate and anthropogenic factors to desertification processes.

4.   The techniques can be applied on a wide variety of temporal and spatial scales. A practical approach would be to select lake & enclosed coastal gulf - watershed eco-systems in degrading and desertified areas for both contemporary and historical monitoring. In the Greek context, the enclosed Gulfs of Elefsis and Amvrakikos on the mainland coast would be suitable and smaller embayments such as those on the Island of Lesbos would also provide appropriate contexts for study. Smaller lake basins elsewhere with representative areas of degradation in their catchments would also be ideal. On a much larger scale, the sediments of the East Mediterranean itself would provide a record of the evolution of the landscape of a whole major region over the last 10,000 years, and this could form the general context within which smaller scale studies could be set.

FIGURE 1   Sediment studies and ecological degradation: empirical
studies, models and management linkages.

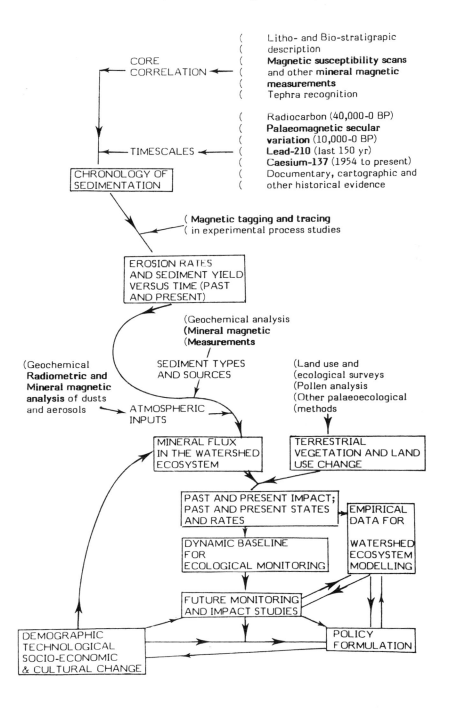

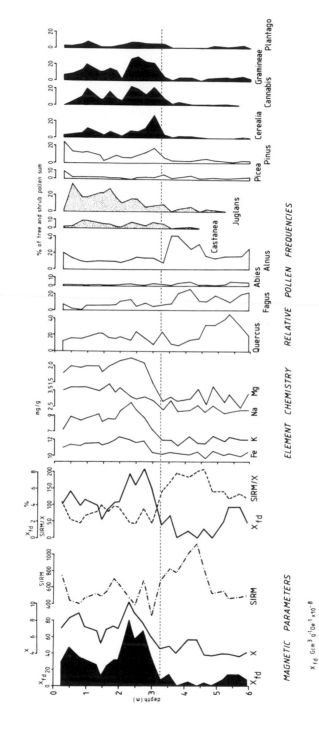

Figure 2 The Record of Human Impact on Lake Sediments from Eastern France

Mineral magnetic, chemical and pollen analyses from the Lac d'Annecy. The depth at which all parameters change dramatically (shown by a dotted line) is dated to c 1100 AD and marks the beginning of intensive medieval farming, for religious houses and for subsistence. Soil loss and chemical output from the catchment accelerate dramatically despite which, on this lithology, productivity is maintained for a long period.

S HIGGITT

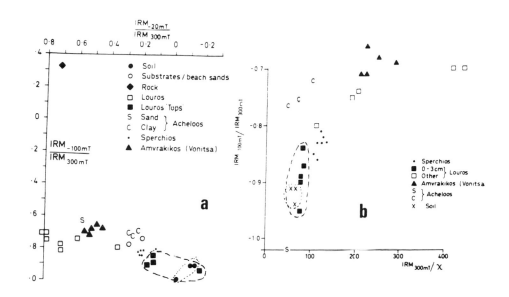

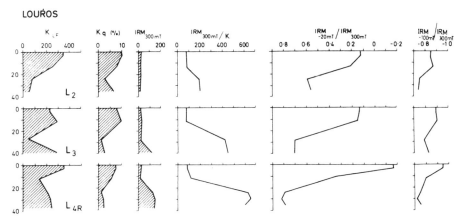

FIGURE 3   Detecting eroded soil in Greek estuarine sediments. Mineral magnetic
measurements from the mouths of the Louros (Amvrakikos Gulf), Sperchios
(Maliakos Gulf) and Acheloos rivers as well as the Amvrakikos
mouth near Vonitsa. Plots (a) and (b) show interparametric ratios.
Plot (c) shows values for samples from shallow (40 cm) cores.
In all cases, the recent Louros samples alone can be identified
as soil derived.

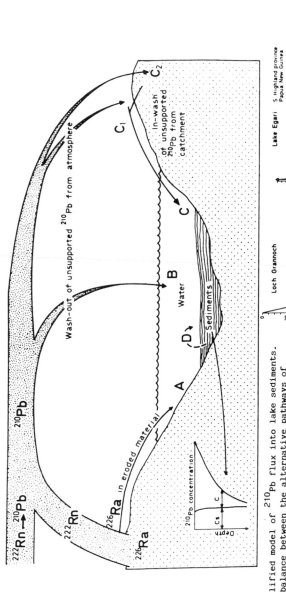

FIGURE 4   Simplified model of ^{210}Pb flux into lake sediments.
The balance between the alternative pathways of
unsupported (excess) ^{210}Pb flux determines the
appropriateness of alternative dating models. The
inset diagram shows the concentration of both
supported and unsupported ^{210}Pb versus depth under
conditions of constant sediment accumulation.

The L. Grannoch plots show the dramatic impact of
commercial afforestation round the lake in 1960. The
L. Egari plots show the impact of subsistence and,
since 1950, western contact on erosion rates at
the site.

REFERENCES

Appleby, P. G. and Oldfield, F. 1983 "The assessment of ^{210}Pb data from sites with varying sediment ccumulation rates". Hydrobiologia, 103, 29-35.

Appleby, P. G., Dearing, J. A. and Oldfield, F. (in press). Magnetic based studies of erosion in a Scottish lake-catchment. I Core chronology and correlations. Limnol. Oceanogr.

Arkell, B., Leeks, M., Newson, M. and Oldfield, F. 1982. Trapping and tracing: some recent observations of supply and transport of coarse sediment from Upland Wales. Spec. Publ. Int. Assoc. Sediment. 6, 117-129.

Arkell, B., Oldfield, F. (In press). "Magnetic tracing of river bedland in the Severn". Proceedings of Conference on 'Temporal studies of erosion and sediment movement: new developments'. Coventry, 1983. Eds. J. A. Dearing & I. Foster.

Bloemendal, J., Oldfield, F., and Thompson, R., 1979. "Magnetic measurements used to assess sediment influx at Llyn Goddionduon". Nature, 280, 50-51.

Bloemendal, J., 1982. Unpublished Ph.D. Thesis, University of Liverpool.

Dearing, J. A., 1983. "Changing patterns of sediment accumulation in a small lake in Scania, South Sweden". Hydrobiologia, 103.

Dearing, J. A., Elner, J. K. and Happey-Wood, C. M. (1981). Recent sediment influx and erosional processes in a Welsh upland lake-catchment based on magnetic susceptibility measurements. Quat. Res. 16, 356-372.

Longmore, M. E., O'Leary, B. M., Rose, C. W. and Chandica, A. L. 1983. Mapping Soil Erosion and Accumulation with the Fallout Isotope Caesium-137. Australian Journal of Soil Research, 211, 373-385.

Mackereth, F. J. H., 1971. "On the variation in direction of the horizontal component remanent magnetization in lake sediments". Earth and Planetary Science Letters, 12, 332-338.

Oldfield, F. 1977. "Lakes and their drainage basins as units of sediment-based ecological study. Prog. Phys. Geogr. 1, 460-504.

Oldfield, F., Rummery, T. A., Thompson, R. and Walling, D. E. 1979. "Identification of suspended sediment sources by means of magnetic measurements; some preliminary results". Water Resources Research. 15, 211-218.

Oldfield, F., Appleby, P. G. and Thompson, R. 1980. "Palaeoecological studies of lakes in the Highlands of Papua New Guinea. I Chronology of sedimentation". J. Ecol. 68: 457-77.

Oldfield, F. 1983. "The role of magnetic studies in palaeohydrology". In "Background to Palaeohydrology; a Perspective". K. J. Gregory, (Ed.). Wileys 141-165.

Oldfield, F. 1983. "Man's impact on the environment". Geography, 149, 2, 167-181.

Oldfield, F., Thompson, R., Dickson, D. P. E. 1981. "Artificial enhancement of stream bedload: a hydrological application of superparamgnetism". Physics of the Earth & Planetary Interiors, (26) 107-124.

Oldfield, F., Battarbee, R. W. & Dearing, J. A. 1983. "New approaches to recent environmental change". The Geographical Journal. 149, no. 2, 167-181.

Oldfield, F., Maher, B., Donoghue, J., Pierce, J., In prep. "Particle-size related, mineral magnetic source-sediment linkages in the Rhode River Catchment, Maryland, U.S.A."

Oldfield, F. and Appleby, P. G. 1984a "Empirical testing of ^{210}Pb dating models for lake sediments". In 'Lake sediments and environmental history'. Eds. E. Y. Haworth and J. W. G. Lund, 93-114. Leicester University Press.

Oldfield, F. and Appleby, P. G. 1984b. "A combined radiometric and mineral magnetic approach to recent geochronology in lakes affected by catchment disturbance and sediment redistribution. Proceedings of Symposium on dating of recent sediments at Sedimentology Congress, Hamilton, Ontario, Aug, 1982. Special publiction volume of Chemical Geology.

Oldfield, F., Appleby, P. G., Worsley, A. T. (In press). "Evidence from lake sediments for recent erosion rates in the Highlands of Papua New Guinea". In B.G.R.G. Symposium Volume on Tropical Erosion. Ed. I. Douglas. George Allen & Unwin.

Oldfield, F., Maher, B. A. 1984. "A mineral magnetic approach to erosion studies" in proceedings of Conference on "Drainage Basin Erosion and Sedimentation". R. Loughron (Ed.) A.N.U. Press 1984.

Oldfield, F. and Appleby, P. G. 1984c. "The role of ^{210}Pb dating in sediment based erosion studies" in proceedings of Conference on "Drainage Basin Erosion and Sedimentation". R. Loughron (Ed.) A.N.U. Press 1984.

Oldfield, F. and Scoullos. Particulate pollution monitoring in the Elefsis Gulf; The Role of Mineral Magnetic Studies. Marine Pollution Bulletin. In Press.

Rummery, T. A. 1983. The use of magnetic measurements in interpreting the fire histories of lake drainage basins. Hydrobiologia, 103, 53-58.

Rummery, T. A. 1981. Unpublished Ph.D. Thesis. University of Liverpool.

Thompson, R., 1973. "Palaeolimnology and Palaeomagnetism". Nature, 242, 182-184.

Thompson, R., Battarbee, R. W., O'Sullivan, P. E., Oldfield, F., 1975. "Magnetic susceptibility of lake sediments". Limnology and Oceanography, 20, 687-698.

Thompson, R., & Oldfield, F. (In press). "Environmental Magnetism". George Allen & Unwin.

Walling, D. E., Peart, M. R., Oldfield, F., & Thompson, R., 1979. "Suspended sediment sources identified by magnetic measurements". Nature, 281, 110-113.

AN ECO-GEOMORPHOLOGICAL APPROACH TO THE SOIL DEGRADATION AND EROSION PROBLEM

A.C.IMESON
Laboratory of Physical Geography and Soil Science,
University of Amsterdam
Dapperstraat 115, 1093 BS Amsterdam, The Netherlands

Summary

The processes of soil degradation associated with desertification are
reviewed. Changes in soil structure occurring when natural ecosystems
are disturbed, are examined with respect to the effect that they have
on erosion and the soil water balance. How soil physical properties
can best be evaluated for erosion control is described using examples
from Mediterranean areas. Some of the difficulties in evaluating the
soil erosion hazard are indicated. For predicting the erosion hazard
it is necessary to understand how ecological, hydrological, pedological and erosional processes are interwoven. For combatting
erosion on cultivated land attention should be given to improving the
soil structure.

1.0 INTRODUCTION

Desertification has been described by Dregne (1978) as "the process
of impoverishment of terrestrial ecosystems under the impact of man. It is
the process of deterioration in those ecosystems that can be measured by
reduced productivity of desirable plants, undesirable alterations in the
biomass and diversity of the micro and macro fauna and flora, accelerated
soil erosion and hazards for human occupancy". Although the effects of
desertification are most dramatic in terms of human tragedy in semi-arid
regions, almost every part of the world is affected. Soil degradation and
erosion are two aspects of the desertification problem which appear to be
increasing in severity in Europe. However, the true extent of soil degradation and erosion within the EEC is unknown. There have been no
systematic surveys and available information is scattered and fragmented.
There are many reasons for this, some economic and political, others due
to the conceptual complexity and multidisciplinary nature of the problem.

In 1977 lack of information about erosion in the United States led to
the passage of the Soil and Water Resources Conservation Act which called
for a detailed survey of U.S. soils (Brown, 1981). In many states alarming-
ly high rates of erosion were reported. An important aspect of the soil
erosion problem is that in areas of excessive erosion government cost
sharing in erosion control measures is necessary because the cost to the
farmer in reducing soil erosion far outweigh the immediate economic
benefits (Brown, 1981).

Surveying soil degradation and erosion are hindered by the lack of an
accepted standardised methodology for evaluating soil physical properties
that indicate the potential for erosion should external factors change.
In this paper the soil degradation problem is firstly reviewed against the
background of changes which occur in soil physical properties when natural
ecosystems are disturbed. Secondly, recommendations are made for evaluating
soil structure for erosional surveys. Finally, the problem of obtaining

information on the soil erosion hazard is discussed and an ecological-
geomorphological approach to this problem outlined. The generalisations
which are described throughout the paper are illustrated by specific
examples from Mediterranean areas.

1.1 The soil degradation process

The causes of soil degradation and erosion in northern and southern
areas of Europe are not the same, but the effects are often very similar.
In semi-arid and Mediterranean areas, the classic example of soil degradation is one in which a reduction in the organic matter content of the
soil results in a decrease in aggregate stability and porosity. Soils
slake upon wetting, soil crusting occurs and infiltration rates are
dramatically lowered. Lower water retention and infiltration rates mean
that less water is available for plants, but that more water is available
for runoff on slopes and in channels. The result is an increasing drought-
iness of the soil, the erosion of the most fertile horizon and a depletion
of the seed bank (Hekstra, 1984). The eroded material causes sedimentation
problems in streams and river channels and disrupts the hydraulic properties
of the river system and the ecology of riparian areas (Swanson et al., 1982).
The capacity of the soil to retain nutrients is reduced and the microclimate of the soil altered. The changes in the water balance of the soil
or slope influences the salt balance with sometimes equally catastrophic
results for erosion. There are many examples from Australia where,
following the clearance of Eucalyptus, a higher seasonal groundwater-table
has developed and salts transported by subsurface flow to slope foot
positions. An increased erodibility of the valley bottom soils due to salt
accumulation has resulted in tunnel or gully erosion and in the occurrence
of salt scalds (Conacher, 1975).
      The onset of soil degradation in seasonally dry climates is usually
attributed to land management practises which have led to a reduction in
the organic matter content of the soil. In dry climates the amount of
organic matter production is low and particularly in Mediterranean regions
highly calcareous soils or high sodium adsorption ratios can make soils
particularly sensitive to a reduction in organic matter. In more humid
areas, aluminium and iron oxides, a preponderance of 2:1 clay minerals,
and acid soil conditions result in inherently more stable soils. The supply
of stabilising organic material to the soil is less subject to fluctuations
resulting from irregular rainfall. Since soil aggregation improves under
some crops and decreases under others (Low, 1972), soil structure can be
managed to produce optimum conditions. However, sometimes economic
conditions favour the repeated cultivation of crops such as maize for
several years in succession. This is the case in the erodible loess soils
of South Limburg in The Netherlands, where erosional problems have become
serious with the recent increase in the area under maize. In Missouri it
has been demonstrated that whereas a rotation of maize, wheat and clover
resulted in an annual soil loss of 2.7 tons/acre/year, comparable land
planted continuously with maize lost 19.7 tons/acre/year (Brown, 1981).
Maize leaves the soil exposed to raindrop impact for relatively long
periods and in the Netherlands excessive amounts of liquid manure applied
as fertilizer result in a deterioration in soil structure. Maize roots
have also been demonstrated to release substances which decrease soil
stability (Reid and Goss, 1981). Molds and algae also seem to stabilise
soil crusts which develop beneath maize and lead to a high frequency of
rill and gully erosion.
      Soil deterioration also results from compaction by increasingly heavy

wheeled farming vehicles. This problem has been studied in the American cornbelt by Voorhees (1979), who found that compaction may actually improve aggregation in the short term, masking negative changes produced by lower organic matter, but in the long term the effects are particularly damaging with respect to root growth in the upper 30 cm of the soil. In South-Limburg in the Netherlands, as in many other areas, gullies and rills are most frequent where compaction occurs at the field boundaries, where tractors turn during ploughing and compaction produces low rates of infiltration. Since soil stabilisation is dependent on soil microfauna, any pollution which inhibits microbiologic activity in the soil has a potentially serious effect on soil aggregation.

2.0 CHANGES IN SOIL STRUCTURE AND INFILTRATION WHEN "NATURAL" ECOSYSTEMS ARE DISTURBED

2.1 Soil Structure

How a soil changes when the vegetation is disturbed is in detail complex and dependent on the parent material and the vegetation. Information is not available for most combinations of vegetation and soil type and the degree to which the generalisations descrbied below are valid still has to be ascertained.

In general, when the organic matter content is > 1%, large water stable soil aggregates are built up of micro aggregates or primary particles, bound together by small roots and fungal hyphae (Tisdall and Oades, 1982). These large aggregates are frequently porous and ensure a high water retention capacity and good aeration. When these aggregates are disrupted, they break down into micro aggregates, some of which are composed of highly stable organic-rich agglomerations 2 to 20 μm in size.

Aggregates bound in this way from forested soils tend not to breakdown when suddenly moistened. In the absence of fine roots and fungal hyphae, aggregates from comparable soils under cultivation, upon rapid wetting, frequently slake into micro aggregates. The stabilising effect of fine roots and fungal hyphae is transient, since this organic matter is continuously being digested and broken down. Organic matter binding micro aggregates, particularly polysaccharides, appears to be more persistent. Removing or reducing the source of organic material reduces the number of water stable aggregates and the smaller structural units that result tend to have lower rates of water acceptance. Under natural vegetation roots and soil fauna not only stabilise aggregates, they also result in aggregate formation. In many non-acid forest and grassland soils almost all of the aggregates may sometimes be formed by fecal pellets.

The size distribution of the breakdown products of large aggregates for comparable soils from cultivated and forest sites are shown in Figure 1. The aggregates, from highly calcareous soils were subject to the impact of falling waterdrops. The forest soils had organic matter contents averaging 7 per cent and the cultivated soils 0.9 per cent. Whereas the forest soils had high rates of water acceptance, the cultivated soils developed crusts and had very low infiltration rates.

2.2 Infiltration

A large population of water stable agrregates gives rise to a low bulk density and high hydraulic conductivity. Frequently, where there is a large soil meso and micro fauna, there is a high macroporosity which gives rise to very high infiltration rates. In the absence of a high

macroporosity a useful method for characterising the water acceptance
characteristics of a soil is by means of the infiltration envelope (Rubin,
1968; Smith, 1972).

The infiltration envelope describes the relationship between rainfall
intensity and the time or amount of rainfall required to pond the soil.
This relationship is more useful than the final steady state infiltration
rate, since the latter is seldom reached under natural conditions. Also
the infiltration envelope can be compared with rainfall intensity
durations and frequencies to calculate the frequency of ponding.
Infiltration envelopes can be determined in the field with a rainfall
simulator (Imeson and Kwaad, 1982) or they can be calculated from measurements of hydraulic conductivity and sorptivity with an infiltration
equation (Smith and Parlange, 1978; Scoging and Thornes, 1980 and Hamilton,
1983). The infiltration envelope at a site is not constant but varies with
the state of soil aggregation and with the soil moisture content.

The effect of a deterioration is soil structure on the infiltration
envelope is illustrated in Figure 2. The most important factor in influencing infiltration was the presence of a surface crust. The magnitude of
the effect of the crust on infiltration can be seen in Figure 3 where
cumulative infiltration rates are plotted against time (mins). The values
were obtained from rainfall simulation experiments during which the soil
was maintained at a ponded condition over 40 % of the target area by
continually varying the simulated rainfall intensity. For comparison values
are included for sites under maquis vegetation and at sites where the soil
has been compacted by sheep and goats. In all cases the infiltration
measurements were obtained during dry antecedental conditions with volu-
metric soil moisture contents in the upper 2 cm < 2 %.

2.3 Erosion as a hydrologic adjustment

Soil erosion can be seen as the effect of an adjustment of the hydrological cycle to ecological change. The hydrologic effect of a deterioration in soil structure is well known. Before considering large changes
such as those caused by crust development, it is important to stress that
even seemingly minor changes in forest ecology can have an impact on the
hydrology and sediment budget of a drainage basin. A good example is
illustrated by the study of old growth Douglas Fir forests on western
slopes of the Cascade Range, reported by Franklin et al. (1981). These
authors concluded that 175 to 250 years are required to develop the old-
growth characteristics of the 350 to 750 year old Douglas Fir - Western
Hemlock forests typical of the Cascade Range. In these old forests the
hydrology, sediment and nutrient budgets are more different than for
younger forests. In the old-growth forests mortality balances growth but
organic matter increases through the accumulation of dead tree boles. The
old growth forests also retain nutrients more efficiently. The small rivers
and streams by which they are drained have channels in which the stability
is maintained by coarse organic debris. This not only prevents erosion but
provides a rich range of habitats and a source of nutrients for stream-
life.

Replacing one type of forest with another can have a great effect on
the runoff and sediment budget. For example, under the ancient mixed oak-
beech forests, the soils on the Keuper marls of Luxembourg are neutral and
support a rich fauna. Many macro pores exist and the movement of water as
throughflow, in the A11 and A12 horizons above an impermable B horizon,
is an important process during wet periods (Bonell et al., 1984). The more
acid soil conditions under introduced conifers result in a reduction in

macroporosity and in subsurface flow so that the size of flood peaks is reduced and the soil is less well drained.

The hydrological cycle operating on a slope is regulated by the vegetation and the soil. Under a vegetation canopy a uniform rainfall distribution tends to be redistributed by the effects of the canopy and by accumulating organic debris, to produce a more variable amount of ground level precipitation. The classic example is the beech forest where soil moisture levels are far higher where stemflow is concentrated near tree trunks and at sites midway between trees, where throughfall is at a maximum.

In semi-arid areas, the process whereby plants intercept water and redirect this as stemflow to their roots has been recently quantified by De Ploey (1982). This process can increase the effective rainfall intensity at the base of grass tussocks by 50 to 100 %. Rainfall beneath a forest tends to have a larger median drop size than rainfall outside of a forest so that its energy is higher. However, in the Colombian Andes it was found that the erosive energy of the rainfall was similar on forested and cleared land in most cases since the higher energy created by the larger drop sizes was compensated for by a higher interception loss (Vis, 1985). Under forest, however, much of the energy may be concentrated at drip points.

The extremes in precipitation amount and intensity that a forest canopy creates usually presents few problems since the infiltration capacity of the soil is generally high. Nevertheless water repellency and drip points can create potentially erosive conditions.

When the process of soil degradation occurs, the soil has usually to accomodate more rainfall, because the opportunities for storage as interception on plant surfaces and residues are less. This requires an adjustment of the hydrological cycle. If infiltration rates remain sufficiently high but less water is stored in the soil and less water is used for transpiration, the lateral movement of water downslope can result in an increased amount of seepage at slope foot positions. Depending on the soil and climate, salinisation, mass movements or gully erosion can occur. If infiltration rates are reduced due, for example, to crusting, and if the microtopography can not retain the ponded water, surface runoff will occur and on steep slopes lead to rilling. Again depending on soil conditions gully erosion may take place. When soil chemical conditions (high sodium adsorption ratios and low electrolyte conditions) favour clay dispersion, infiltration rates are particularly low (Table I) and the sediment concentrations in runoff high. Much of the sediment transported by runoff is deposited as colluvium at lower hillslope positions and the rest in river channels. Colluvial deposits are often layered and have a depositional surface crust. In semi-arid areas they are frequently the first materials that develop ponding and surface runoff.

Just as the erosion that occurs on hillslopes is an adjustment to the changing water retention and infiltration characteristics of the soil, so also is the erosion which takes place in river channels. These have to adjust to accomodate higher peak discharges and sediment loads. Since this process can take very many years and because different channel forms become stable under the new conditions, flooding, sedimentation and the erosion of irrigated land by streambank processes become the norm. Particularly where metastable pedisediments or alluvial deposits have accumulated during more stable recent geological time, arroyo type gully erosion may occur.

The implications of considering soil erosion as a hydrological adjustment to altered ecological conditions, will be considered later.

3.0 EVALUATING SOIL PHYSICAL PROPERTIES

The susceptibility of a soil to erosion depends on not only its physical characteristics but also its position in the landscape, vegetational effects, the microclimate and the type of erosion process being considered. It is unrealistic to consider erodibility as a function of soil physical properties alone but it is nevertheless necessary to be able to evaluate erodibility as a function of relatively easily measured or observed soil characteristics or properties. This has proved very difficult and there are still no standardised procedures that have been internationally accepted. One difficulty is the diversity of soil type and the fact that different soil properties account for the stability of different soils. The most widely used erodibility index is the "k" factor of the Universal Soil Loss Equation (Wischmeier, 1978) which can be used for estimating soil loss by surface wash processes over long periods of time. The k factor was obtained from statistical relationships developed in the United States and can be approximated with a nomogram requiring soil texture and organic matter parameters. It has not been possible to verify its applicability to European conditions due to a lack of measured soil loss data.

Simple methods of evaluating soil physical properties which indicate erodibility need to be tested under European conditions. Perhaps the best indication of erodibility is provided by field measurements of infiltration but these have not been widely applied, probably because they are more time consuming than other methods. Examples of infiltration techniques have already been given (Figure 2 and Figure 3). These results were obtained with simple field portable rainfall simulators which are easy to construct and which require little water. Useful classification schemes have been developed to indicate the water stability of soil aggregates. These are based on tests which involve recording the reaction of aggregates when they are emersed in water with or without simple pretreatments (Emerson, 1967; Loveday and Pyle, 1973). Also useful are tests which involve determining the water dispersible clay or silt fractions after various pretreatments. Useful procedures have been developed in Australia (Chittleborough, 1982; Shanmuganathan and Oades, 1982) and used in England (Reid and Goss, 1981) and Spain (Imeson and Verstraten, 1985). The amount of dispersible clay in particular has been demonstrated to be related to soil physical properties associated with erodibility and soil degradation.

The water stability of soil aggregates can be measured in the laboratory by wet-sieving, water drop or ultrasonic dispersion techniques. Examples of obtained results with this type of tests are indicated in Figure 4 and Table II. A promising index is the consistency index of De Ploey and Mücher (1981) which has been demonstrated to be related to crust development in loamy soils.

The greatest research need is for field measurements of soil loss which can be used to evaluate the effectiveness of easily determined aggregate stability tests. Given such data it should be possible to characterise soil erodibility rapidly and easily in the future for soil survey purposes.

For tunnel and gully erosion different soil properties influence the erodibility besides those mentioned above. The properties which are important vary with the type of gully being considered (Table III; Imeson et al., 1982).

4.0 THE EROSION HAZARD

4.1 Erosional phenomena and geomorphological perspectives

The occurrence of erosional phenomena, such as rills, gullies, truncated soils and colluvial deposits, can be readily mapped from air-photos or field surveys. It must be stressed, however, that such phenomena indicate more the state of erosion than the present rate of soil loss or the potential for further erosion. In some cases erosional phenomena do provide indications of trends in stability and a warning of the potential erosion hazard, but such interpretations can be misleading without an appreciation of geomorphological and ecological conditions.

It is an obvious point that the erosional phenomena in a landscape are formed by past erosional events and that their persistance indicates the relative rates of their formation and obliteration. Where infrequent rainfall events are most important and of restricted areal extent, as in semi-arid regions, the marks of erosion persist for hundreds or maybe thousands of years, creating a landscape which at first appears to be suffering from rapid and severe soil erosion but which in fact is relatively inactive. This "moon crater effect" appears to be the case with badland landscapes in Israel, for example, which although having apparently fresh erosional features, have hardly changed since prehistoric times (Yair et al., 1980). This is probably also true for the Rio Fardes badlands in southern Spain, but not for all badlands, some of which are actively eroding today (Bryan and Campbell, 1982).

Another complication is the distinction between natural and accelerated erosion. Erosional phenomena associated with accelerated erosion in more humid areas may be natural phenomena during periods of ecological stress in more arid regions. Langbein and Schumm (1958) and later Kirkby (1980) have considered how sediment yields are influenced by climate on a continental scale. Under natural vegetation, maximum sediment yields occur in areas of about 300-500 mm rainfall because, according to the authors, at lower rainfall amounts the amount of runoff that can transport sediments becomes a limiting factor and at higher precipitation levels, the effects of vegetation reduce the erosivity of the rainfall and promote a better soil structure. However, whilst it is true that sediment yields of rivers are lower in semi-arid areas, this may be partly because material is only transported over short distances and does not reach the major river channels. In the Rif mountains of Morocco for example it was found that in an area receiving 300 mm rainfall per year, although severe erosion was occurring on slopes, much of the eroded material was being deposited on pediments or in wadis where transmission losses are high. The acute soil loss and sedimentation problems were not reflected in the sediment yields of the drainage basins.

An important consideration in interpreting erosional phenomena concerns the existance of geomorphic thresholds (Schumm, 1979). Erosion usually occurs when threshold conditions are transgressed by an extreme event. Identifying threshold conditions is a useful means of locating areas of potential erosion hazard. Schumm (1979) and Graf (1979) have applied this approach in the western United States to pinpoint valley bottoms where arroyo type gully incision is a potential hazard. They could do this simply by identifying threshold stream discharge or energy and valley gradient conditions required for arroyo development.

Related to the threshold consideration is the difficulty that a) some time (the reaction time) may elapse between the disturbance leading to erosion, and the onset of erosion, and b) time will elapse between the onset of erosion and the achievement of a steady state (relaxation time; Figure 5). This problem has been described for gully development by Graf (1977) for the Colorado Piedmont in the neighbourhood of Denver. It is apparent that a potentially erodible situation can not always be identified by erosional phenomena (which are not yet present). Also, once erosional features such as gullies or truncated soils are present, the adjustment to the disturbance may have reached a new stable equilibrium and there may be no further erosion or erosion hazard. This is certainly the case in very many parts of Europe. A prerequisite for erosion is erodible soil or weathering products. Once these have been lost they are generally replenished at such slow rates that further erosion at the same site can only occur under unusual circumstances.

4.2 Evaluating the erosion hazard

Prior to any evaluation of potential erosion hazard the geomorphological context of the erosion problem needs to be established. From geomorphological and pedological surveys it is frequently possible to isolate areas of potential erosion hazard where research should be concentrated. This research should be geared to investigating the effects of clearly specified ecological changes, not on erosion in general but on particular erosional processes and transport mechanisms. Since hydrological processes operate on slopes and drainage basins, it is necessary to consider changes at this scale. This is useful because soil properties frequently have a catenal sequence which, once understood, can be used for mapping potential soil erodibility. Hydrological adjustments at a site can not be examined properly without reference to the effects of simultaneously occurring adjustments at upslope and downslope sites.
The requirements for such an approach include:
1) Clearly defined objectives and description of ecological changes,
2) Geomorphological and pedological surveys to indicate location of potentially unstable soils,
3) An evaluation of the erosional effects of hydrological adjustments in response to the ecological changes.
A major objective should be to make suggestions for minimising the erosion hazard by earmarking certain sensitive areas for conservation, or by suggesting ways of preventing or ameliorating a deterioration in soil structure.

Rates of soil loss

The problem of measuring or calculating rates of soil loss has not been described. The Universal Soil Loss Equation (Wischmeier and Smith, 1978) has frequently been used in Europe for estimating actual rates of soil loss. Recently, simulation models such as CREAMS (Knissel, undated) have been developed for predicting runoff, erosion and chemical losses from field soil sites. The CREAMS model has been tested on a few occasions in Europe, but there is not enough experience to evaluate its usefulness.

5.0 RECOMMENDATIONS

The general problem

The soil degradation and erosion problem in Europe requires evaluation in terms of its historic, present and potential extent. This evaluation should include an economic appraisal with respect to production loss on agricultural land and flooding and sedimentation damage along rivers. The qualitative damage to (semi)natural ecosystems is not easy to quantify economically, but it is potentially serious, particularly if such areas are to serve for recreation.

The soil degradation problem has to be tackled at the level of the drainage basin since the problem involves the hydrologic cycle and the movement of water and nutrients along various pathways in drainage basins. At the drainage basin level, carefully planned water and soil conservation programmes are required which take account of local soil and climatic conditions and the requirements of the population.

Experience elsewhere has shown that halting soil erosion is expensive and often uneconomic on agricultural land, when set against short term benefits. Financial or Community assistance is often necessary to protect the essentially non-renewable soil resource. Many soil and water conservation schemes have been initiated for example, in the United States and Australia and these have been evaluated in detail at various levels of decision making. This experience from outside of Europe should be evaluated carefully with respect to its possible relevance for the European problem.

Specific Research Requirements

There has been an underinvestment in soil erosion research in Europe. That research which is being done requires co-ordination and common objectives. These could usefully include:
1) Development and standardisation of techniques for evaluating soil physical properties associated with erodibility. Attention should be given to methods which can be combined with soil surveys and to the problem of soil crust development.
2) Establishing stations where erosion rates can be measured over long periods to provide basic data comparable with that available in the United States. This data should be used to test mathematical models and techniques referred to under 1).
3) Research into the processes of soil structure deterioration occurring through a) pollution, b) disturbance of (semi)natural ecosystems.
4) The designation of special protection zones which should have restrictions designed to preserve the ecological environment.
5) Establishing a data bank of information on soil erosion.

The above recommendations are not original nor exhaustive. They are put forward as a basis for further discussion.

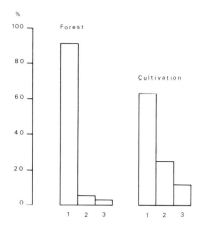

Figure 1   Size of breakdown products of large aggregates subject to splash erosion. Average values are given for ten forested and ten cultivated sites in the neighbourhood of Teulada, Southeast Spain. (Class 1 = > 2.8 mm; class 2 = 0.21 to 2.8 mm; class 3 = < 0.21 mm).

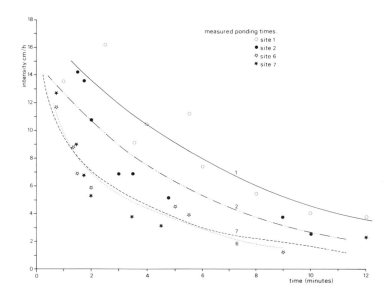

Figure 2   Infiltration envelopes for erodible soils from Northern Morocco. Site 1 is for a red Mediterranean soil recently cultivated and having a good structure. Site 7 is for a similar soil with a surface crust. Similarly sites 2 and 6 are for non-crusted and crusted regosols.

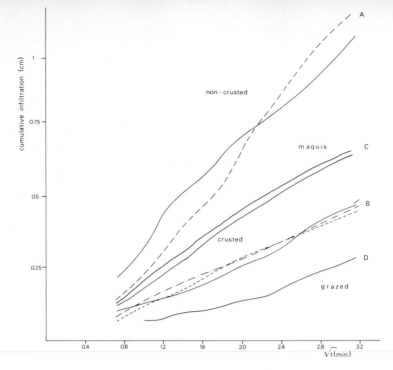

Figure 3   Cumulative infiltration rates measured with a rainfall
            simulator in Northern Morocco: A. cultivated non-crusted
            Mediterranean soil; B. same soil with crust; C. maquis;
            D. red soil grazed and trampled by goats and sheep.

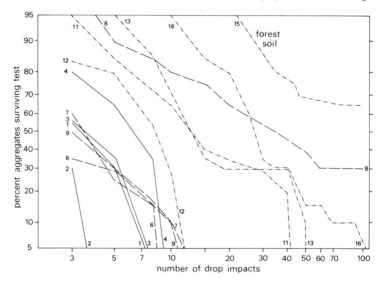

Figure 4   Aggregate stability of highly calcareous soils from South-
            east Spain, as indicated by waterdrop test: soils 1 to 9
            are from badland sites; samples 6, 7 and 9 are saline soils;
            samples 11, 12 and 13 are cultivated regosols from Teulada
            (Alicante), sample 16 is a red soil from the same location
            and sample 15 is a soil from a forest site (from Imeson and
            Verstraten, 1985).

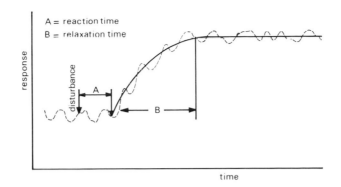

Figure 5   Response of a geomorphic system subject to disruption
                  (based on Graf, 1977).

Table I. Example of rainfall simulation experiment results.
         Time (t) and amount (P) of precipitation required
         to produce overland flow at various locations in the
         area of Alicun (Granada, Spain).
         All sites are highly calcareous.

LOCATION	INTENSITY 8 cm.h^{-1}		INTENSITY 4 cm.h^{-1}	
	t (min)	P (mm)	t (min)	P (mm)
Valley bottom				
1. ploughed	2.9	3.8	10.42	7.3
2. crusted	1.8	2.3	6.25	4.4
Badland pediment (high ESP)				
crust	1.2	1.6	4.3	3.0
crust	1.8	2.3	6.2	4.4
ploughed	3.1	4.1	11.1	7.8
Alluvial fan				
crust	1.0	1.3	3.51	2.5
crust	2.3	2.9	7.9	5.5
soil with shrinkage cracks	4.9	6.4	17.35	12.1

Table II. Results of aggregate stability determinations for Mediterranean
soils using the waterdrop test and the dispersion index (DI) of
Loveday and Pyle (1973). For the waterdrop test the average
number of drop impacts required to break down the tested
aggregates is indicated. Values of DI range from 0 (no dispersion)
to 16 (complete dispersion).

	Average number of drop impacts (range)	DI (range)
Northern Morocco (Beni Boufrah)		
A. steep stoney cultivated slopes	10–38	0–3
B. maquis	45–73	2–4
C. red Mediterranean soil subject to slaking	11–18	0–5
D. alluvial fan (colluvial red soil)	18–25	3–5
E. saline soils	12–55	0–15
F. pine forest	> 75	3–4
G. "duplex" profile	5–30	0–14
Southern Spain		
A. badlands Guadix	3–5	2–10
B. highly calcareous forest soils	100+	1
C. highly calcareous cultivated regosols	8–15	0–2
D. saline soil (gully site)	1–2	10

Table III. Sets of conditions characterising particular gully types (Imeson and Kwaad, 1980)

Gully type	Gully cross section	Position in landscape	Principal source of runoff	Material in which gully is developed	Favourable conditions
Type 1	V-shaped	Anywhere, except valley bottoms, where runoff becomes concentrated	Overland flow[1]	Relatively resistant weathering products of impermeable parent materials, B horizons of deep soil profiles. The resistance of the material does not decrease with depth	Intense rainfall, poor soil structure, steep slopes, poorly built terraces and tracks
Type 2	U-shaped	Anywhere in landscape, except valley bottoms	Overland flow[1], with a contribution of subsurface water of lesser importance; occasional seep caves at headcut	Relatively little resistant weathering products or slope deposits which do not increase in resistance with depth.	Dispersive soil materials, sub-humid climate with pronounced wet and dry seasons.
Type 3	U-shaped	Anywhere, but usually on pediments and gentle lower slopes	Subsurface flow predominates, as is apparent from piping	Weathering products and slope deposits as for type 2	Dispersive soil materials are essential; further as type 2
Type 4 (arroyos)	U-shaped	Valley bottoms	Overland flow[1], mainly from tributary gullies, and subsurface flow	Alluvial and slope deposits	Semi-arid climate, lack of valley bottom vegetation, dispersive materials

1) In addition to overland flow produced by the areas between gullies, by terraces, roads etc., runoff may be produced in the gully itself due to the low infiltration capacity of the gully floor (Yair et al., 1980).

REFERENCES

Bonell, M., M.R.Hendriks, A.C.Imeson and L.Hazelhoff, 1984. The generation of storm runoff in a forested clayey drainage basin in Luxembourg. Journal of Hydrology, Vol. 71, 53-77.

Brown, L.R., 1981. Eroding the base of civilisation. Journal of Soil and Water Conservation, Vol. 36, 255-260.

Bryan, R.B. and I.A.Campbell, 1982. Surface flow and erosional processes in semi-arid meso-scale channels and drainage basins. Intern. Assoc. Sci. Hydrol. Publ., 137, 123-133.

Chittelborough, D.J., 1982. Effect of the method of dispersion on the yield of clay and fine clay. Australian Journ. of Soil Research, Vol. 20, 339-346.

Conacher, A.J., 1975. Throughflow as a mechanism responsible for excessive soil salinisation in non-irrigated, previously arable lands in the Western Australian wheat belt: a field study. Catena, Vol. 2, 31-67.

Dregne, H.E., 1978. Desertification: Man's abuse of the land. Journal of Soil and Water Conservation, Vol. 33, 11-14.

Emerson, W.W., 1967. A classification of soil aggregates based on their coherence in water. Australian Journ. of Soil Research, Vol. 5, 47-57.

Franklin, J.F., K.Cromack Jr., W.Denison, A.McKee, C.Maser, J.Sedell, F.Swanson and G.Juday, 1981. Ecological characteristics of old-growth Douglas Fir forests. USDA For. Serv. General Techn. Rep. PNW-118, 48 pp. Pacific Northwest Forest and Range Experiment Station, Portland, Oregon.

Graf, W.L., 1977. The rate law in fluvial geomorphology. American Journ. of Science, Vol. 277, 178-191.

Graf, W.L., 1979. The development of montane arroyos and gullies. Earth Surface Processes, Vol. 4, 1-14.

Hamilton, G.J., I.White, B.E.Clothier, D.E.Smiles and I.J.Packer, 1983. The prediction of time to ponding of constant intensity rainfall. Journal of Soil Conservation, New South Wales, Vol. 39, 188-198.

Hekstra, ., 1984. Paper to be presented at the Conference.

Imeson, A.C. and F.J.P.M.Kwaad, 1980. Gully types and gully prediction. Geografisch Tijdschrift Vol. 14, 430-441.

Imeson, A.C. and F.J.P.M.Kwaad, 1982. Field measurements of infiltration in the Rif Mountains of Morocco. Studia Geomorphologica Balcanica, Krakow.

Imeson, A.C., F.J.P.M.Kwaad and J.M.Verstraten, 1982. The relationship of soil physical and chemical properties to the development of badlands in Morocco, 44-70 in: Bryan, R.B. and Yair, A. (eds.). Badland Geomorphology, GeoBooks, Norwich, 408 pp.

Imeson, A.C. and J.M.Verstraten, 1985. The erodibility of highly calcareous materials from southern Spain (in press).

Kirkby, M.J., 1980. "The Problem". Chapter 1 (pp. 1-16) in "Soil Erosion" (ed. M.J.Kirkby and R.P.C.Morgan, J.Wiley & Sons, Chichester, 312 pp.

Knisel, W.G., undated. Editor CREAMS. A field-scale model for chemicals, runoff and erosion from Agricultural Management systems. U.S.Dept. of Agriculture Conservation Research Report no. 26, 640 pp.

Langbein, W.B. and S.A.Schumm, 1958. Yield of sediment in relation to mean annual precipitation. Transactions American Geophys.Union, Vol. 39, 1976-1084.

Loveday, J. and J.Pyle, 1973. The Emerson dispersion test and its relationship to hydraulic conductivity. Division of Soils Techn. Paper 15, 1-7. CSIRO Australia.

Low, A.J., 1972. The effect of cultivation on the structure and other

characteristics of grassland and arable soils (1945-1970). Journal of
Soil Science, Vol. 23, 363-380.
Ploey, J.de, 1982. A stemflow equation for grasses and similar vegetation.
Catena, Vol. 9, 139-152.
Ploey, J.de, and H.J.Mücher, 1981. A consistency index and rainwash
mechanisms on Belgian loamy soils. Earth Surface Processes and
Landforms, Vol. 6, 319-330.
Reid, J.B. and M.J.Goss, 1981. Effect of living roots of different plant
species on the aggregate stability of two arable soils. Journal of
Soil Science, Vol. 32, 521-541.
Rubin, J., 1968. Numerical analysis of ponded rainfall infiltration.
Internat. Assoc. of Scientific Hydrology, Symposium Wageningen,
pp. 440-451.
Schumm, S.A., 1979. Geomorphic thresholds: the concept and its applications.
Inst. of British Geographers Transactions, New Series, Vol. 4, 485-515.
Scoging, H.and J.B.Thornes, 1980. Infiltration characteristics in a semi-
arid environment. Internat. Assoc. of Scientific Hydrology, Pub. 128,
159-168.
Shanmuganathan, R.T. and J.M.Oades, 1982. Effect of dispersible clay on
the physical properties of the B horizon of a red-brown earth.
Australian Journal of Soil Research, Vol. 20, 315-324.
Smith, R.E., 1972. The infiltration envelope: results from a theoretical
infiltrometer. Journal of Hydrology, Vol. 17, 1-21.
Smith, R.E. and J.Y.Parlange, 1978. A parameter efficient hydrologic
infiltration model. Water Resources Research, Vol. 14, 533-538.

Swanson, F.J., R.L.Fredrikson and F.M.McCorison, 1982. Material transfer
in a Western Oregon Forested Watershed, p. 233-266 in: Edmonds, R.L.
(ed.): Analysis of coniferous forest ecosystems in the Western United
States. US/IBP Synthesis Series 14, Stroudsburg, Pa. Hutchinson Ross
Publishing Co.
Tisdall, J.M. and J.M.Oades, 1982. Organic matter and water stable
aggregates in soils. Journal of Soil Science, 33, 141-163.
Vis, M., 1985. Rainfall characteristics and splash erosion in forest eco-
systems of the Central Colombian Andes (in prep).
Voorhees, W.B., 1979. Soil tilth deterioration in the northern Corn Belt:
Influence of tillage and wheel traffic. Journal of Soil and Water
Conservation, Vol. 34, 184-186.
Wischmeier, N.H. and D.D.Smith, 1978. Predicting rainfall erosion losses -
a guide to conservation planning. U.S. Dept. of Agriculture,
Handbook no. 537, 58 pp.
Yair, A., H.Lavee, R.B.Bryan and E.Adair, 1980. Runoff and erosion
processes and rates in the Zin Valley Badlands, Northern Negev,
Israel. Earth Surface Processes, Vol. 5, 205-225.

HAZARD MAPPING AS A TOOL FOR LANDSLIDE PREVENTION IN MEDITERRANEAN AREAS

Th.W.J. Van Asch
Department of Physical Geography, Geographical Institute
State University Utrecht

Summary

In the Mediterranean area the degradation of the landscape by
landsliding is a serious problem. The climatological conditions,
with a strong concentration of rain in the winter period is one of
the main causes of landslide activity in these areas. Also the
strong tectonic and neoteconic activities have led to a high
frequency of mass movements. Hazard mapping is a good tool for the
prevention of these landslides because these maps contain
information about actually or potentially hazardous areas. The maps
should be used in the early stage of a planning activity for future
economic development of an area. It is suggested that the different
types of hazard maps has to be critically evaluated for certain
areas and knowledge and experience on this matter has to be
exchanged between different countries.

1. INTRODUCTION

In the Mediterranean area the degradation of the landscape by
landsliding is one of the most serious forms of landdegradation in time
and space. These slope instability phenomena cause a serious obstacle in
the economic development of these areas. In South Italy, Carrara et al.
reported that nearly 18% of the area is unstable (6). Twenty percent of
all the events registrated in this area produced dangers to all types of
structure while a far greater percentage of events affected arable land.
Coumantakis and Angelidis reported that in Greece, agriculture in the
mountaineous areas is mainly concentrated on unstable land (13). From
1950-1980, 60% of all the sliding events affected the agricultural
villages.
All sorts of technical measures have been developed for the
stabilization of landslides. However, the application of these techniques
is very expensive. Therefore in areas where landslides occur very
frequently, more attention has to be given to landslide prevention. A
great part of preventive measures can already be taken in the early
planning stage of a certain area, by using so-called hazard maps. The
hazard or risk maps give information about the degree of risk for
landsliding within different land units of a certain region. There is an
urgent need for the production of risk maps which can be easily read by
non-specialists, who are involved in all sorts of planning procedures. In
countries like France and Italy legislative activities have been
developed in order to oblige the consultation of hazard maps before
planning a construction. The risk maps should give clear statements about
the degree of risk for landsliding in different zones. It is the purpose
of this paper to discuss different types of hazard maps with regard to
their suitability for risk prevention and landslide prevention.

2. FACTORS RESPONSIBLE FOR LANDSLIDING IN MEDITERRANEAN AREAS

2.1 The process of landsliding
Many types of landsliding can be distinguished (24). It is usual to classify landslides on the basis of:
1. the mechanism of failure and transport of material
2. the type of material involved in landsliding
3. the amount of displacement with time (velocity)
4. the activity of the slides
5. the size of the slides.

The equilibrium of a slope mass is determined by the ratio between the shear stresses (T) which tend to set the mass into motion and the shear strengths of the materials (S) which form a resistance against the applied stress. The safety factor which is a measure for the degree of stabilitiy is defined as:

$$F = \frac{S}{T} \geqslant 1$$

A slope is considered as stable if the safety factor is greater than 1.
There is a number of primary factors which determine this safety or stability factor. These factors are:
1. the strength parameters c and φ
2. the bulk density and depth of the material
3. slope angle and height of the slope
4. the maximum pore water pressure conditions.
There is a large number of landscape factors (secondary factors) which influence directly or indirectly the above mentioned parameters. Factors which are important for slope stability problems in the Mediterranean countries are shown in Fig. 1.

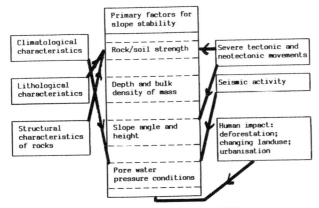

Fig. 1. Mediterranean factors causing slope instability.

2.2 The climatological factors

One of the most important climatological characteristics which favour landsliding in the Mediterranean areas is the precipitation, which can be relatively high, and is mainly concentrated in the winter periods. This concentration of rainfall gives rise to high pore water pressures in the slope mass during the winter periods and this forms the main trigger for the landslide activity in these areas (5, 7, 13, 20, 27). In the high mountains of the Southern French Alps it is the combination of snowmelt and heavy precipitation in spring which can create dangerous situations (Tricart, 1974).

2.3 The geological factors

The geological factors which favour the susceptibility for landsliding in these areas are:
1. the dominance of clay-material especially from Tertiary and Quaternary age
2. the weak consolidation of many Tertiary deposits
3. the intensive fracturing of the rock materials due to strong neotectonic activities during the Tertiary and Quaternary periods
4. the plastic characteristics of many Paleozoic rocks
5. the presence of a thick cover of altered periglacial and glacial material on slopes consisting of impermeable rocks.

Systematic investigations by Carrara et al. carried out in Southern Italy revealed that on a regional level the lithological characteristics are the most dominant agents determining the intensity of landsliding (7, 8). Especially the rocks of Tertiary age with a strong clay component are susceptible to landsliding (see also Van Asch, 1980). Other rock types which are in various ways susceptible for mass movements are the gneisses, schists and phyllites of Paleozoic and Mesozoic age, the marls and sandstones and flysh deposits of Tertiary age. Also it appeared that in these areas the intensity of fracturing and faulting is an independent variable which can influence strongly the intensity of mass movements (7). In the Pyrenees there are e.g. the plastic Paleozoic rocks and the Triassic and Jurassic limestones which slide over the Triassic marls and account for many landsliding events (9, 12, 25). Coumantakis and Angelidis mentioned two important lithological units in the mountaineous areas of Greece which are very susceptible for mass movements: the molasses and the heavily fractured and folded flysh deposits (13). In the Alpine regions especially the Jurassic and Quaternary clay deposits and the morainic deposits show many mass movements.

2.4 Seismic activity

A very important trigger for landslide activity in these Mediterranean areas are the seismic activities. These seismic shocks are responsible for many large landslide events which affected many towns and accounted for a lot of victims. Coumantakis and Angelidis mentioned that 50% of all European seismic activity is concentrated in Greece (13).

Seismic shocks increase the stress conditions in the slope mass and decreases the intergranular friction in the soil mass. This may cause liquifaction in especially sandy material, which decreases the bearing capacity of the ground. This produces complete subsidence of buildings and other types of artificial constructions (3, 23). Also dormant deep seated landslides can be reactivated by the seismic shocks which bring about a deformation of the earth surface over a large area. This has resulted in some cases in severe urban damages (14).

2.5 Slope angle

Many parts of the mountaineous regions in the Mediterranean countries were strongly uplifted during the Pleistocene period. The rapid incision of rivers has led to the formation of steep slopes. On these slopes mass movements are the dominant erosion process. The relationship between slope angle and intensity of landsliding, however, is in detail a rather complex one. In Southern Italy there is an overall similar pattern, in which landslides incidences are bound to a lower but also to an upper limiting slope angle. Depending on certain rock types it can be shown that the percentage of instable terrain increases with increasing slope angle towards a maximum and then decreases with increasing slope angle (Carrara et al., 1982).

Van Asch showed on the basis of geotechnical considerations, that the frequency of landslide events increases with slope angle but the volume of landslide material per event decreases with increasing slope angle (27). Therefore slope angle cannot be simply introduced as a dominant parameter influencing the intensity of landsliding.

2.5 Vegetation and landuse factors

It has been said that deforestation in the Mediterranean countries has led to an increase of the unstable area. Therefore it seems appropriate to focus our interest on aforestation activities because these activities can be one of the most important remedial measures for the stabilization of slopes over large areas. However, there is a gap in our knowledge about the precise effect of different types of vegetation on the stability of slopes in these areas. This will be one of the main topics of research for the future in the department of Physical Geography in Utrecht. Carrara points to the fact that steep slopes with a natural vegetation (forest) cover are in general more stable than deforestated slopes (7). Van Asch doubted, however, whether on flatter slopes with more deep seated landslides a forest cover can lead to stabilisation (27). Also Carrara et al. mentioned that aforestation measures had not until now improved the stability of the area (6). The effect of forest vegetation on the stability of slopes is rather complex. Fig. 3 gives a schematical picture of the influence of relevant vegetation factors on the degree of stability of a slope. The roots of the vegetation cover increase the soil strength and therefore, especially in shallow

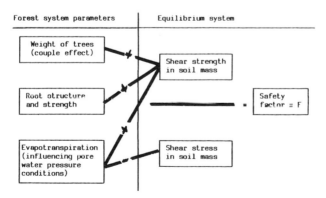

Fig. 2. The influence of vegetation factors on the degree of stability of the slope.

regoliths, the stability of the slopes. The evapotranspiration of a dense forest cover determines the maximum pore water pressure conditions in the soil. It has been shown that cutting down of a forest cover results in an increase in the maximum water pressures and in a decrease in the stability of the slope (32). The weight of the trees on the soil increases the soil strength but also the shear stresses in the soil. Especially on flatter slopes with wet cohesive material an increasing weight of the trees can result in a failure of the slope (18). An increase in the overburden pressure by the trees may cause creep processes (slow flow processes) in the soil (4). Also surface shears and moments due to wind effect on especially large trees are serious threads for the stability of the slopes (17). At this moment in the French Alps special grass vegetation is used for the stabilization of the slope in order to reduce the negative effect of the weight of the trees on the stability.

A change in the agricultural landuse can also have a great influence on the stability of the slopes. In the French Alps the change into a new large scale agricultural landuse structure, has led to a neglect of the maintenance of the old drainage systems. This caused an acceleration in the frequncy of landslide incidences during the last 15 years in this area (1).

3. HAZARD MAPPING: THE MAIN TOOL FOR LANDSLIDE PREVENTION

There exists a large variety of remedial measures which can stabilize landslides (aforestation, drainage, loading of the toe and unloading of the head, walls and dams) but these measures are very expensive. It is by far more convenient and less expensive to avoid unstable slopes and this can only be done if all those areas which are actually or potentially hazardous are mapped out in the very early stages of any planning (8). Different types of hazard maps have been developed for this purpose (11, 31). According to Varnes a landslide hazard or stability map gives a division of the land surface in homogeneous land units and the order of degrees of actual/potential hazard due to mass movement (31). An ideal risk map should contain information about:
a. the soil and rock mechanical parameters
b. the depth of the materials
c. the maximum pore water conditions
d. the slope topography (28).
With these parameters one is able to calculate the degree of stability of a certain slope or land unit. However, it is extremely difficult to collect the above mentioned parameters over a larger area. Therefore most hazard maps are constructed on the basis of other factors. The general concept behind the construction of these maps is the general geological principle that "the past and present are keys to the future" (8). This means that slope failures in the future wil be more likely to occur in those land units where the geological, geomorphological and hydrological conditions have led and will lead to landslide activity in the past and present. It is obvious that the prediction of risks in landslide maps is still very difficult because of a lot of varying secondary factors, mappable and not mappable, which determine the stability of the slopes. Fig. 3 gives a survey of different types of hazard maps which have been developed for different scales.

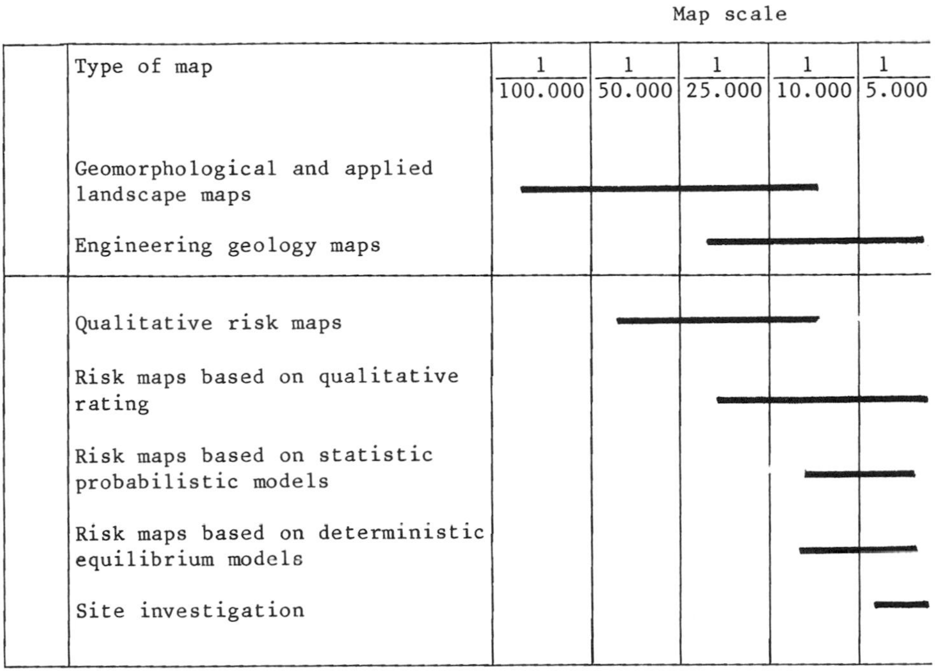

Map scale

Type of map	$\frac{1}{100.000}$	$\frac{1}{50.000}$	$\frac{1}{25.000}$	$\frac{1}{10.000}$	$\frac{1}{5.000}$
Geomorphological and applied landscape maps		▬▬▬	▬▬▬		
Engineering geology maps			▬▬▬	▬▬▬	▬
Qualitative risk maps		▬▬▬	▬		
Risk maps based on qualitative rating			▬▬	▬▬▬	▬
Risk maps based on statistic probabilistic models				▬▬	▬▬
Risk maps based on deterministic equilibrium models				▬▬	▬▬
Site investigation					▬▬

Fig. 3. A survey of different types of hazard maps.

3.1 Engineering geological maps

In the so-called engineering geological maps special interest is focussed on the geotechnical characteristics of the materials, which in most cases are presented in combination with the registration of the landslide phenomena. Classification schemes have been set up, which enable to describe some properties of soil and rock material in the field, related to the strength characteristics of the material (26). Direct information about the strength of rocks and soil material can be obtained by simple measurements of strength in the field carried out in as many places as posisble or by back analysis on landslides (28). Sissakian et al. give a good example of different scales of an engineering geological map for a region in the West-Pyrenees (25). The main item, given in colours, is the difference in strength of the rock and soil material. The total strength of the rock has two components: the unconfined compression strength of the sound rock (measured with the point load test or "Hammerschmidt" test) and the spacing of joints in the rock. These parameters can be easily determinated in the field. The soil (loose material) is classified according to the USCS soil classification which gives qualitative information about the soil mechanical behaviour of the material.

The map of Fenti et al. gives, apart from the strength of the rock material also an indication of the main directions of the joints in the rocks, which is also very important for the prediction of the stability of the slopes (16). Also the map of e.g. Centamore et al. gives geotechnical information of the rock and soil material (10). On these maps other important factors are given like the thickness of the soil material, hydrological conditions, landuse and vegetation and anthropic activities.

In most cases the engineering geological maps give geotechnical information about the materials in combination with the different types of landsliding. They can only be interpreted by scientists specialized in the stability problems of slopes.

3.2 Classification of risk zones based on qualitative terrain analysis

One of the basic principles in making (qualitative) risk maps is based on the fact that the systematic recording of geomorphological phenomena can be used as a basis for an evaluation of regional differences in slope stability. Further it is assumed that landunits which are classified according to rock material and structure, slope angle and vegetation are comparable with regard to the risk of damage by mass movements (27). With this assumption in mind the observed intensity of mass movement in well defined land units can be used to evaluate the potential danger of slope instability for parts of the area that are not yet affected by some kind of instability (30). A good example of these types of maps are the so-called ZERMOS maps (1:25.000) which are developed for various regions in France. These coloured maps show the degree of risk for landsliding for a certain area. Special attention for these maps is given to those areas which are not yet affected by landslides but which can be very dangerous in the future, because they have the same landscape characteristics as certain areas which are already unstable.

Dumas et al. proposed a legend for a detailed qualitative risk map (scale 1:10.000) for a region in Calabria (15). The classification of units according to the degree of risks (in colours) gives information about:

a. the severity of events especially with regard to human activities

b. the probability of occurrence of the events.

Van Steijn and Van den Hof found that a more detailed observation about landsliding processes in relation to slope angle, materials etc. can result in a finer differentiation of the degree of risk than given by the ZERMOS maps (30). They showed that a map printed in black and white, can give very clearly and in a much cheaper way information about the degree of stability in certain regions. Worth mentioning is also the hazard map developed by Bocquet et al. for mountaineous (Alpine) areas, in which the relation between natural risk processes and degree of human activity in different land units are integrated (2).

A good example of assessing the risk of landsliding in a more systematic and objective way is proposed by Kienholz (19). By means of a checklist procedure and rating systems the elaborator is obliged to determine systematically the degree of hazard. The maps varying in scale from 1:25.000 to 1:5.000 are suitable for different planning purposes. The terrain has to be subdivided into units according to topography, lithology, landuse and slope aspects and on the basis of landslide types which are classified pragmatically according to their size and degree of activity. Five forms must give all the information about historical data, detailed environmental factors, climatological characteristics, etc. The rating of this information results in a final assessment of the degree of stability of a certain unit.

3.3 Hazard maps based on statistical analysis of the landscape variables

In the above given examples of hazard mapping a high degree of subjectivity is involved in the weighing and rating of the different slope instability factors and in assessing the different hazard levels. Therefore computer aided statistical techniques have been applied by several investigators in an attempt to analyse quantitatively the relationship between certain geological features of landslide masses and the slopes where they occur (8, 21, 22). Different statistical methods were tested by Carrara on material which was collected during 6 years of investigations in different key-areas in southern Italy (8).

By means of discriminant analysis Carrara (1983) tested a great number of geological and geomorphological variables for their discriminant power in predicting whether certain zones are stable or unstable. The standardized discriminant coefficients of these variables can yield useful information concerning the factors controling slope instability of these sample areas. The classification function scores, which can readily be converted to probabilities of sliding, were then used to make 4 classes of landslide hazard probability. Also the percentage of unstable area per unit area was regressed using multiple regression analysis. The multiple regression model for the area proves to be a fairly good model for predicting landslide affected slope units. Carrara suggested that strong negative deviation between the predicted degree of instability and the real instability in a certain unit, which means that there is in reality less instability than predicted, may indicate a strong degree of susceptibility for future landsliding (8). A positive deviation between predicted stability and observed stability may point to an exhausted or stabilized landslide zone.

There are certain drawbacks for this way of investigation:
1. there is relatively much time and money involved for setting up the data basis
2. the same attributes contribute in a very different way to the degree of instability for different regions.

Therefore generalizations of the influence of variables is very difficult. Therefore a regionalization of different types of areas with the same geological and geomorphological conditions seems necessary (7).

3.4 Hazard mapping using deterministic geotechnical equilibrium models

It is striking that very few attempts have been made to use a number of existing geotechniccal equilibrium models, describing the degree of stability (F-value) of a slope in order to determine in a deterministic way the risk for landsliding on a regional scale. We already mentioned before that it is not easy to find the required parameters for the stability model. However, Van Asch showed that the determination of the strength parameters of the materials did not necessarily have to be carried out by laborious and expensive measurements in the laboratory (28). Detailed mapping of a number of landslides, from which the hydrological conditions and the geometry of the slip surface are known can give an estimate of the mean and variance in strength and the depth of a certain type of material. The variety in strength parameters and depth of the material can be used to calculate the range in the degree of stability, expressed as F-values, of a certain land unit (see Figure 4). If we assume that the variety of the values of geotechnical parameters are normally distributed within a certain landunit it is also possible to calcultate the probability of landsliding within this land unit (29). One

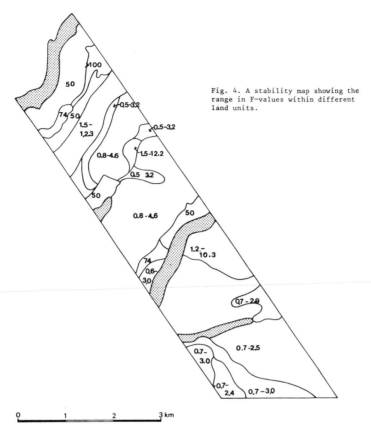

Fig. 4. A stability map showing the range in F-values within different land units.

of the problems which has to be solved more accurately is the determination of the variety of pore water pressure conditions. This can be done by a study of the precipitation history and the infiltration characteristics of the soil (32). In this way risk maps can be constructed on a scale 1:5000 to 1:10.000 which are based on a deterministic approach of the stability problem.

4. RECOMMENDATIONS FOR THE EEC RESEARCH PROGRAM

Landsliding is one of the main agents of degradation of the landscape in the Mediterranean areas. Investigations have to be carried out on an international scale in order to detect those areas, where slope instability phenomena cause a serious obstacle in the economical development. Hazard mapping on a regional scale is a good method for the prevention of further landsliding because in this way an impression of the areal extent and the degree of instability is obtained. The hazard maps with different scales can be used during early planning procedures for the economic development of a region. These maps are also necessary for the planning of remedial measures.

The different systems of hazard mapping has to be critically evaluated for certain areas and knowledge and experience on this matter has to be exchanged between different countries. There has to come an improvement of the quality and quantity of basic data. New mappable environmental factors (especially hydrological factors) which are well correlated to slope instability should be sought. For this purpose the use of remote sensing techniques has to be evaluated.

More attention has to be given to those types of hazard mapping which are based on statistical probability models and geotechnical equilibrium models.

Research has to be stimulated with respect to the influence of different vegetation types on the stability of the slopes in order to find the best remedial measures for slope stabilization on a regional scale.

The effect of the change in agricultural structures on slope stability in a region has to be studied carefully in order to improve landuse planning.

REFERENCES

1. BESSON, L. (1984). Service de restauration des terrains en Montagne de L'O.N.F. Direction Departementale de l'Agriculture de L'Isère. Colloque mouvements de terrain. Université Caen.
2. BOCQUET, F., CHARRE, J.P., F. DOUARD, J.L., THOURET, J.C., VIVIAN, H. (1984). Carte integrée des risques naturels et anthropiques en milieu montagnard (Notice explicative). Colloque mouvements de terrain. Université Caen.
3. BOUSQUET, B. and PECHOUX, Y. (1984). Les mouvements de terrain facteur géomorphologique d'interêt regional en Grèce Centrale. Colloque mouvements de terrain. Université Caen.
4. BROWN, E.T. and SHEWS, M.S. (1975). Effect of deforestation on slopes Journal of the Geotechnical Engineering division, ASCE, 96 SM6 1917-1934.
5. BRUM-FERRERA, A. de (1984). Mouvements de terrain dans la région au nord de Lisbonne. Colloque mouvements de terrain. Université Caen.
6. CARRARA, A., PUGLIESE-CARRATELLI, E. and L. MERENDA (1977). Computor based data bank and statistical analysis of slope instability phenomena. Zeitschrift für Geomorphologie N.F. 21(2), 187-222.
7. CARRARA, A., SORRISO-VALVO, M. and C. REALI (1982). Analysis of landslide form and incidence by statistical techniques, Southern Italy. Catena 9( ), 35-62.
8. CARRARA, A. (1983). Geomathematical assessment of regional landslide hazard. Fourth International Conference on Applications of Statistics and Probability in Soil and Structural Engineering 4-27.Université de Firenze. Pitagora.
9. CAZENAVE-PIARROT, F., LAUGENIE, C., TIMAY, J.P. and R. BOURROULIK, (1984). Controle géologique et morphoclimatique des glissements de versants dans les Pyrenees occidentales. Colloque mouvements de terrain. Université Caen.
10. CENTAMORE, E., CHERUBINI, C. EUSEBIO, L.D., DRAMIS, F. GENTILI, B. MARCHETTI, P. and F. PONTONI, (1981). Cartografia geomorfologica a indirizzo application: un esempio nell'area marchigiana. Bollettino dell'HIC 53 11-15.
11. COTECCHIA, V. (1978). Systematic reconnaisance mapping and registration of slope movements. International Association of Engineering Geologists Bulletin 17, 5-37
12. COROMINAS, J. and E. ALONSO, (1984). Inestabilidad de Laderas el Pirineo Catalian. Tipologia y Causas. Jornadas de Trabajo sobre Inestabilidad de Laderas en el Pirineo. Barcelona, Universidad Politechnica de Barcelona.
13. COUMANTAKIS, J. and CH. ANGELIDIS, (1984). Mouvements de terrain en Grèce; aspects socio - economiques. Colloque mouvements de terrain. Université Caen.
14. CRESCENTI, U., DRAMIS, F., GENTILI, B. and A. PRATURLON,(1984). The Bisaccia landslide: a case of deep seated gravitational movement reactivated by earthquake. Colloque mouvements de terrain Université Caen.
15. DUMAS, B., GUEREMY, P., LHENAFF, R. amd J. RAFFY, (1984). Risques de Mouvements de terrain dans une region seismique: la facade Calabraise du Detroit de Messine aux abords de Villa San Giovanni (Italy). Méditerranée (in press).

16. FENTI, V., RUZZIER, D., SILVANO, S. and V. SPAGNA, (1981). I movimenti fanosi della Valle Isarco tra Bolzano e Ponte Gardena (Alto Adige) CNRS publication nr. 69, 130 pp. Padova, Instituto de Geologia Applicata.
17. GRAY, D.H., (1970). Effects of forest clear cutting on the stability of natural slopes. Bulletin of the Association of Engineering Geologists, 7, M1-M2; 45-66.
18. GRAY, D.H. and W.F. MEGAHAN, (1981). Forest vegetation removal and slope stability in the Idaho Batholith. U.S. Department of Agriculture; Forest Service Research Paper INI-217.
19. KIENHOLZ, H., (1983). Landslide hazard assesment for landslide zonation. Berne, Geographical Institute University of Berne.
20. LOYE-PILOT, M., (1984). Coulées boueuses en Corse: examples de mouvements de terrain en pays Mediterranéen montagnard. Colloque mouvements de terrain. Université Caen.
21. NEULAND, H., (1976). A prediction model of landslides. Catena 3; 215-230.
22. REGER, J.P., (1979). Discriminant analysis as possible tool in landslide investigations. Earth Surface Processes 4; 267-273.
23. SAURET, B., (1984). Manifestations d'Instabilité du sol dansd la zone epicentrale du seisme de Messme de 1908. Le role de la liquifaction. Colloque mouvements de terrain. Université Caen.
24. SCHUSTER, R.L. and R.J. KRIZEK, (1978). Landslides and control. Special Report 176. 234 pp. Washington, National Academy of Sciences
25. SISSAKIAN, V., SOETERS, R.and N. RENGERS, (1983). Engineering geological mapping from aerial photographs: the influence of photo scale in map quality and the use of stereo orthophotographs. ITC-journal 1983-2, 109-118.
26. U.S., BUREAU OF RECLAMATION, (1974). Earth Manual. Washington US Govt. Printing Office.
27. VAN ASCH, Th.W.J., (1980). Water erosion on slopes and landsliding in a Mediterranean landscape. Utrechtse Geografische Studies 20; 237 pp. Utrecht, Geografgisch Instituut.
28. VAN ASCH, Th.W.J., (1984). Landslides: the deducation of strength parameters of materials from equilibrium analysis. Catena 11; 39-49.
29. VAN ASCH, Th.W.J. (in prep.). Hazard mapping using deterministic models.
30. VAN STEIJN, H. and G.J.J. VAN DEN HOF, (1983). Stability of slopes near Barcelonette (Alpes de Hautes Provence, France): A case study in slope stability mapping. Geologie en Mijnbouw; 677-682.
31. VARNES, P.J., (1982). The principles and practice of landslide hazard zonation. Commission on landslides and other Mass-Movements- IAEG. The Unesco Press, Paris.
32. WU, T.H., SWANTSON, D.N., (1980). Risk of landslides in shallow Soils and its relation to Clearcutting in Souteastern Alaska Forst Science, 26 (3), 495-510.

THE "LUCDEME" PROGRAM IN THE SOUTHEAST OF SPAIN TO COMBAT DESERTIFICATION IN THE MEDITERRANEAN REGION

A. PEREZ-SOBA and F. BARRIENTOS

Instituto Nacional para la Conservación de la Naturaleza (ICONA)

Summary

The National Institute for the Conservation of Nature (Ministry of Agriculture of Spain) has established the LUCDEME program (Lucha contra la Desertificación en el Mediterráneo = Desertification control in the Mediterranea) in the southeastern area of Spain (30.000 Km^2) where the most eroded lands and desertificated areas of the country are located. The main objectives of this project are: a) to achieve a better understanding of the desertification process, causes and effects, b) to improve applied technologies to combat desertification: past and present activities in the area, design of new technologies, c) education, training and extension in the field of desertification.
Several universities, national institutes and other national and regional bodies are involved in this program, the contribution of wich will substantially help to improve the understanding of the problems involved and to design new approaches for their solution.

1. Background.

The long process of deterioration of natural resources in Mediterranean countries has been caused by human action, wich has affected an especially fragile natural environment through inappropriate use of land and resources.
This fragility is principally the result of low, irregular rainfall which limits the development of the vegetation and hinders the recovery of the ecosystems altered by man. The more rugged the topography of natural spaces and the more susceptible they are to successive violent torrential phenomena, so much more serious the situation become.
The combined action of all these causes has led to erosion in large areas of Mediterranean countries. In its turn, the progressive exhaustion of natural resources has led to the socio-economic decline of the rural areas affected. Human communities end up suffering the grave consequences. All too often through lack of knowledge, the lack of valid alternatives, or through attempts to extract yields incompatible with the conservation of resources, humans degrade the natural environment where they settle.
For many years, science and technology have been concerned with partial aspects of desertification, and recently the whole phenomenon has come under study at an international level, witness the 1977 World Conference on Desertification. This conference pointed out the necessity of advances in the knowledge of all the factors that lead to desertification -physical, economic, demographic, sociological, etc.- and also ot the consequences in each of these areas.

Emphasizing the possibilities offered by international collaboration, this conference, after recommending that national action should be taken, stressed the importance of sub-regional cooperation in this field.

As can be seen from the papers presented at the conference and also from the World Desertification Map, Spain presents great problems in this field. Its situation in the Mediterraean and man's traditional activities affecting the environment have been the principal causes of the present situation. This could become considerably worse as a consequence of the process of development. Therefore, for many years, there has been much activity aimed, above all, at the restoration of renewable natural resources.

Exemples of this on a large scale are the reafforestation programmes and the activities on watershed management and torrent control.

The Spanish government is interested in promoting cooperation among the Mediterranean nations on desertification. It believes that the work already done and the projects in progress can serve the interest of Mediterranean countries. Therefore, Spain has taken the initiative to put forward the program described below.

In order to study the problem of desertification, the program will try to isolate the different factors which are involved in the process.

It is considered that at one point in the development of the program it will be possible to test an integral management model.

To be able to do this, it will probably be necessary to adopt succesive approximations, i.e. moving from relatively simple schemes with few variables, to other more complicated ones and finally to the full model with all its components.

A model of this type does not only try to establish a rational physical proceedure for soil conservation. The economic, demographic and social aspects are also very important. It is thus a "human" model, because what is being aimed at is not simply an increase in income levels.

In effect, what is aimed at is a scheme to cover an area big enough for the problems to be considered on a real scale. Then a possible solution may be found to the problems of development/underdevelopment, which can be seen on all scales throughout the world, global, continental, regional, national and even local, thereby motivating general interest towards methodology.

2. Aims and objectives of the program.

2.1. Aims of the program
The program aims to achieve:

Control of desertification in the Mediterranean area.

The program established in accordance with the ideas of the United Nations Conference on Desertification (Nairobi, Kenya 1977) considers desertification as the diminishing or destruction of the biological potential of the earth, or, the process of deterioration of soil and water in conditions of ecological stress.

This definition implies that in the area, qualitatively or quantatively unsuitable activities are or have been carried out.

These activities may be the result of lack of knowledge or experience or of possible alternatives. They may also be due to attempts to obtain large short-term gains at the expense of long-term productivity.

In any case, desertification is normally a slow process. The final solution is also long-term, resulting from education, social and economic progress and the appropriate balance between demography and resources.

The possibilies of short-term action are based on restoration and improvement of land use.

These ideas were established in Nairobi to preface the recommendations of the Plan of Action, approved by the Conference and then ratified by the General Assambly of the United Nations. They are thus the basic coordinates of this program.

The scheme proposed for Spain, and examined in the light of the Nairobi recommendation on international cooperation especially at regional level was the driving factor behind this program.

In fact, conditions similar to those in the area selected may be found all round the Mediterranean basin and comparable situations are found in many countries with a Mediterranean climate in South America and other parts of the world. Therefore, the conclusions to be drawn here may have a far wider application in the future.

2.2. Scope of the Program

As stated above, the area most affected by desertification, and therefore of greatest interest, is the S.E. of the peninsula.

Two criteria, administrative and ecological, have been used in this definition.

Administratively the scope is the three provinces of Murcia, Almeria and Granada. In this way questions such as statistics can be resolved as most of the data is classified on a provincial basis. Contacts with the authorities, public bodies and the people affected are obviously also easier at a provincial level.

The ecological criterion is based on the Mediterranean climate as recommended at Nairobi. According to these recommendations, analyses, studies and actions should always be based on the hydrological basin.

With these three provinces, the basins draining into the Mediterranean will be specially considered.

Therefore the greater part of the province of Murcia, the whole of Almeria and the southern part of the provincie of Granada are within the scope of the project.

The administrative area is 32.622 Km2 and the ecological area 22.597 Km^2.

The population is 1.817.121 and about 1.217.445 respectively.

Among the river basins the following should be mentioned either for their size or their special problems:
- Guadalentin or Sangonera (tributary of the River Segura)
- Almanzora.
- Albuñol.
- Guadalfeo.

2.3. State of desertification of the area

The area of south-eastern Spain selected for the development of this LUCDEME program has been subjected to intense human pressure for generations.

Since ancient times, this area has enjoyed wide markets for its agricultural and mining products, due to its position in the Mediterranean. This has given rise to a market economy wich has been responsible for intensive productive pressure on the environment. This, together with a marked increase in population, has profoundly disturbed the natural ecosystems, causing one of the longest desertification processes known. The extent and depth of deforestation of the mountains is notable; all trace of a fertile horizon has disappeared from the slopes.

This is a rugged area, rising in the Sierra Nevada to over 3000 metres above sea level at distances of only 40 km from the sea. Limy or degradable lixiviate soils are common and the climates range from arid to subhumid. Agriculture is carried out sometimes on areas with a rainfall of less than 200 mm. and the centuries old tradition of sheep farming has put great strain on the fragile ecosystems. The process of desertification is principally the result or erosion by water. The destruction of plant cover in the mountainous areas has led to an increase in direct surface run-off, thus aggravating the natural irregularity of the rainfall with its frequent flash floods and catastrophic consequences.

A study of the state of erosion in the area yields the following data.

-Little or no sheet erosion and stable drainage
system 10%
-Considerable sheet and rill erosion, drainage
system slightly torrential 28%
-Considerable erosion in rills and gullies; -
drainage system torrential 42%
-Considerable erosion in rills and gullies; -
drainage system high torrential 17%
-Areas with land-slides 2%

The following figures show the amount of soil lost through water erosion.

Soil losses t/ha. per year	% of surface
10	24,0
10-25	30,9
25-50	11,2
50-100	26,3
100-200	6,9
200	0,6

From this we see that 76% of the land suffers greater than acceptable soil losses, the average for the area being 42,9 t/ha per year. Soil losses are higher on cultivated slopes, sometimes exceeding 50% and no conservation work at all is carried out on these.

2.4. Objectives of the Program

AIM	MEDIUM TERM OBJECTIVE	IMMEDIATE OBJECTIVES
Control of desertification in the Mediterranean area. (Area proposed for the program: Provinces of Granada, Murcia and Almeria, especially their slopes to the Mediterranean)	I Analysis of the various resources and factors involved in the processes of desertification.	I-1 To obtain knowledge of existing natural resources, their evolution, level of degradation and its causes. Human resources and their evolution. I-2 To discover the basic shortages which the processes of degra-

dation have intro-
duced into the
natural environment
and the influences
of these processes
on the socio-economic
decline of the natu-
ral areas affected.

II Determination of
systems and
techniques wich
can be applied to
combat deserti-
fication.

II-1 Comparison and
improvement of
systems and tech-
niques to combat
desertification.

II-2 To determine the
real and social costs
of restoring natural
resources necessary
to achieve develop-
ment in the rural
areas at high
standards of living.

II-3 Methodological defi-
nition and experimen-
tation to calculate
profitability with
special reference
to mathematical –
models.

II-4 Design of methods
to follow up the
systems and techni-
ques applied.

II-5 Establishment if
possible of integra-
ted planning for
areas affected by
desertification.

III Education,
training and
extension on
the project's
content.

III-1 Technical edu-
cation and trai-
ning programmes.

III-2 Extension pro-
grammes.

IV Anti-desertifi-
cation combat
activities.

IV-1 Restoration of
soils/vegetation.

IV-2 Correction of the
effects of torrential
phenomena.

3. Action in progress.

Action to restore flood torrent courses has been taken since the beginning of the century by the various forestry organizations concerned. Since that time, special interest has been shown in the area where, at least in its first phase, the LUCDEME program will operate; that is the provinces of Almeria, Granada and Murcia.

This activity can be summarized as follows:
- Almeria province: 109,351 ha. of hydrological forest restoration afforestation and 21 torrential basins undergoing correction work.
- Granada province: 23,500 ha. of hydrological forest restoration afforestation and 21 torrential basins undergoing correction work.
- Murcia province: 66,798 ha. of hydrological forest restoration afforestation and 32 torrential basins undergoing correction work.

At present the National Institute for Nature Conservation (ICONA), the national executive organization for the LUCDEME program, is supervising these operations. They are part of its subprogramme "the combating of erosion, destabilization of soils and the loss of water resources". In the decade 1980-1989, the following programme is planned for the LUCDEME area:
- Almeria province: 73,999 ha. of hydrological-forest restoration afforestation and 399 km of torrential watercourse correction.
- Granada province: 57,142 ha. of hydrological-forest restoration afforestation and 484 km of torrential watercourse correction.
- Murcia province : 47,388 ha. of hydrological-forest restoration afforestation and 1,047 km of torrential watercourse correction.

4. Participating organizations.

In order to rapidly achieve the ends hitherto set forth, via the Ministry of Agriculture, working in conjunction with ICONA, specialized cooperation is essential and, hence, various organizations will be involved.

These organizations include universities, technical schools, institutes of scientific investigation together with government technical bodies concerned with agriculture, hydrology, geology, meteorology, etc.

Special mention deserve:
- Instituto Nacional de Investigaciones Agrarias (INIA).
- Dirección General de Extensión y Capacitación Agraria.
- Centro de Estudios Hidrográficos.
- Confederaciones Hidrográficas del Sur y del Segura.
- Universidad Politécnica de Madrid. Escuela Técnica Superior de Ingenieros de Montes.
- Universidad Politécnica de Córdoba. Escuela Técnica Superior de Ingenieros Agrónomos.
- Universidad de Granada. Facultades de Ciencias Biológicas, Geografía, Geología y Farmacia.
- Universidad de Murcia. Facultades de Ciencias Biológicas, Geológicas y Geografía.

- Instituto Geológico y Minero de España.
- Instituto Nacional de Meteorología.
- Consejo Superior de Investigaciones Científicas (CSIC)

Participation of International Bodies or Work Groups specialized in desert projects is expected.

5. Studies and investigations already underway in the program development

Within the programme outlined, various studies and investigations have already been carried out or are in the process of being so.
- Methodological investigation to estimate hydric erosion in small and medium size basins.
- Geographic models applied to soil conservation.
- Definition of soil loss models, in basins, due to surface run-off, slumping and linear erosion in river beds and flood terrain.
- Definition and sediment transport model adjustment along the hydrographic network to a reservoir.
- Study of co-efficient R, rain factor in USLE model, on the mediterranean basin, through digital analysis of pluviographic levels and posterior automatic processing of data. Isoline cartography and obtaining of regional regression for pluviometric values available.
- Study of erosive terrain in South Eastern Spain through automatic treatment of soil data, morphology and plant coverage. Study of control and correspondence of such erosive terrain with LANDSAT information.
- Studies of erosion and surface run-off in torrential basins using conventional mathematical models of soil loss and specific degradation. Checking and adjustment of parameters and systematic use of model for quantitative evaluation of rainfall in mediterranean basins.
- Study of the influence of hydrological-forestry restoration work, on the Southern slope of the Gador Sierra, on the refilling of the Campo de Dalias aquiferous layer, together with the bringing up to date of the hydric balance according to the different plant coverage.
- Qualitative and quantitative evaluation of wind erosion in the Program area.
- The drawing-up of a map of soils in the Program area, with special analysis of state of degradation.
- Pilot study of the physical environment in the Adra river basin, as a methodological outline, for the evaluation of factors and parameters in the desertification process in the South-East.
- Zoological and parasitological aspects affecting degradation of grazing areas in the Program zone.
- Methodological trial to evaluate the state of soil degradation in a representative zone of the South-East, through analysis of substrat micro-organism population.
- Hydric, meteorization and erosion balance in a microbasin, representative of the mediterranean holm oak groves.
- Study of the biological quality of the Guadalhorce river water, through an analysis of the macro-invertebrate population.
- Study of human influences in the desertification process of the River Adra basin.
- Study of the structure and functioning of the drainage network in the Puentes feeder reservoir in the River Guadalentin.

Final considerations:

The above, apart from the importance and peculiarity of the area chosen for the Program, reveals the firm resolve of Spanish authorities to concentrate a great amount of effort in the South-East of the Peninsula, to be applied in the forthcoming years. Given the monumental undertaking implied and the complex nature of desertification and the technical, social and economic resources needed to combat it, this will essentially depend upon the availability of personnel qualified in the various disciplines involved.

This in itself already constitutes an important contribution towards the spread of knowledge and applied technology and could have even greater consequences should the program acquire an international dimension.

Such the program would allow, that the total efforts of those experts on the threshold of current knowledge be combined in one 'workshop', a permanent pool of new ideas and formulae immediately applicable in one given area, where important tasks in process could be concentrated upon.

In short, should the program be managed in this fashion, it would achieve the triple objective of knowing more, acting more efficiently, and making others a party to higher knowledge. Bodies involved in the program, would then have performed a noteworthy service in the worldwide struggle against Desertification.

Further, however, should the model of organized co-operation, here outlined as a prime objective of the program, be achieved, the service could well transcend the specific problem of desertification to cover the problems of the developed-underdeveloped communities, in the widest possible sense. And these are problems wich, ever more acute and worrying, affect the whole planet.

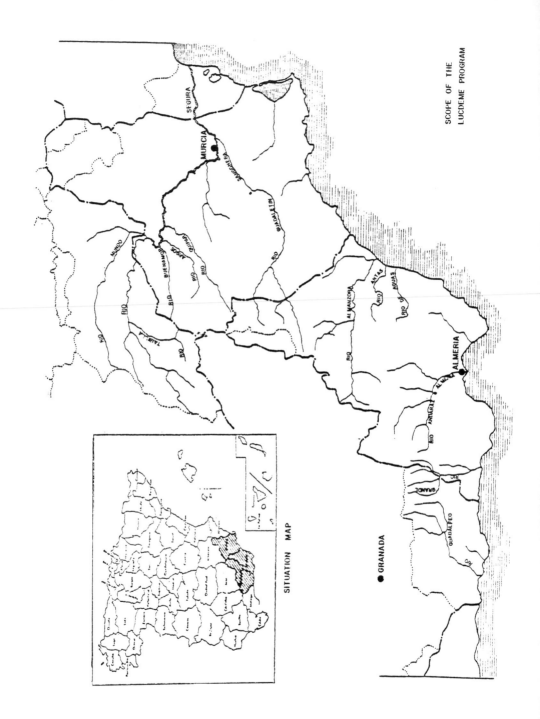

SITUATION MAP

SOIL CONSERVATION PROBLEMS IN ITALY
AFTER THE COUNCIL OF RESEARCH FINALIZED PROJECT

F. MANCINI

Institute of Geopedology and applied Geology
of the University of Florence, Italy

Summary

The Author briefly describes the purposes of the italian C.N.R. finalized project for soil conservation (1976-82). From this experience he deduces which are the main lines to be followed in the battle against the various aspects of the desertification in Europe. The scientific community is ready and well prepared but what we need is a strong pressure of the people to politicians and decision makers to obtain sufficient funds and personnel for this battle against the desertification.

1 - It is well known that Italy has only 20% of its surface that is flat terrain, the residual 80% is hilly and mountainous. Alpine regions have cold climate and abundant snow fall while in the South and in the islands a mild mediterranean climate prevails.

Geolithology is a splendid pallet and consequently morphologies (landscapes) and soils vary continuously.

Soil conservation in such a country is therefore a permanent and delicate problem that all administrations, national, regional and local have to face.

There is a long and important scientific and technical tradition in approaching the deferce of the more or less steep slopes. The concept that the "watershed" is the fundamental study and project unit was established already by the foresters and the hydraulic engineers in the last century.

The National Council of Research of Italy proposed and the governement financed a quinquennial finalized project for "Soil Conservation" at the end of 1976.

This project came to an end in december 1982. Which were the purposes of such a Project and which have been the results, the achievements?

The main tasks of the Project were:

1 - Setting up of new study methods
2 - Appliyng these new methodologies in typical watersheds, experimental areas, characteristics regions
3 - Collection, analysis, procession and filyng of important basic data
4 - Transfer of the more interesting results to the different users
5 - Cooperation and assistance to state and regional agencies working in

the land planning and protection.

It has been for all of us and especially for me that I had the privilege to direct this project a splendid and stimulating experience.

Scientists and technicians showed a highly appreciable cooperative and civic spirit.

The scientific results are included in the 896 publications of the Project. A general Catalogue of these papers has been printed and distributed; it is in italian and english.

2 - Of course soil conservation studies and works did not begin with the overmentioned project nor come now to an end. A lot of serious problems still are in front of us. Among these let us quote: landslides and erosion in the mountains and hills, sediment transport in rivers and little streams, flooding and subsidence in the recent alluvial plains, loss of fertile soils by wild urbanization, drawing back of coasts etc.

Examples of these phenomena are well known and detailed descriptions of exceptional events have been published also recently and in many cases illustrated in international meetings by italian scientists pertaining to the University or to the National Council of Research (see fig. 1 in the next page).

I am very grateful to the colleagues of various europeans countries that in many occasions and also in this meeting have illustrated some of the more interesting results of our researches.

I believe that there are not great difficulties for the technical solution of many of these questions. The obstacles are of different nature.

First of all we lack a recent general law. The forestal act is more than sixty years old (n° 3267 of 1923), other laws (integral rehabilitation 1933, for the mountains various acts 1950-1960) do not include a modern general scheme of soil conservation practices.

We need urgently such a law that must also give guidelines for the regional administrations and their relative legislation.

In my opinion for bringing to a solution the soil conservation problems it is necessary a better coordination among the national, regional and local bureaus and agencies. It is easy to demonstrate that the Italians, as many other mediterraneans, are not leaders in coordination. Indipendence is certainly a right but individualism is often a big danger.

Beside this we can agree that the following three aspects are fundamental, that is:
1 - a well prepared national project
2 - a sufficient amount of money rationally distributed
3 - good technicians without stinginess in quantity.

There is no need of many words to comment the before mentioned three points. Concerning the third one let us assure that a large availability of fresh energies, of well prepared young technicians is at our disposal already from now.

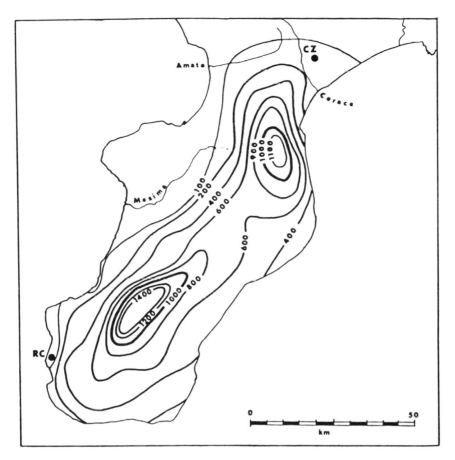

Fig. 1

An exceptional event in Calabria: the precipitations of the 16, 17, 18 October 1951 (from Caloiero and Mercuri, 1980).

It is rather funny that while a diffused unemployment exists there are not initiatives for enlarging the staff of the State and regional agencies dealing with these problems. Only two or three regions, as an exception, are moving in this direction.

Financing is certainly an other important aspect but in Italy is in general much easier to get miliards for an important work, a fundamental structure, but very difficult after that to find the few millions necessary to the maintenance of such realizations.

3 - Before concluding these short commentary on the italian situation in this field it is probably useful to give an idea of which should be the trend of the researches and the contributions to land planning in the near future.

The problem has at least two quite different aspects.

1) Basic research on soil degradation due to water erosion, wood-fires, landslides, flooding must be continued with increasing intensity. In some cases, as will be illustrated in this meeting by A. Aru for Sardinia, these phenomena are so severe that we can speak of "desertification". In many others the productivity is decreasing and the soil losses are still important but overall we have to control possible future dangers.

2) The other aspect has practical implications in land planning. It is the writer's opinion that we have to produce for the benefit of the different users very clear and simple documents. Two points seem important to stress. The legend of the various maps, the explanations of diagrams and nomograms must not include terrible and complicated scientific terms (calcilutites, fragiudalfs, pelloxererts). This is a language to be used among specialists, in strict privacy. If we speak in this way to the users we will surely dissaude them instead of convincing.

Simple and understandable words and clear subdivisions are most suitable. If we use the adjectives "suitable" "moderately suitable" "unsuitable" or "stable" "moderately stable" "unstable" everybody will understand and the decision-makers will fully profit of our work.

The second point concerns what kind of documents we have to produce. Should we give an exact inventory of a given phenomenon or better indicate the existence and the degree of risk, the presence of a danger?

In many of our mountains and hills in every part of Italy there are geolithological formations, for instance the "argille scagliose" mica-schists, phyllades or the varicolour clays, the mio-pliocenic silty clays that are very unstable and erodible. Not everywhere gully-erosion (calan-chi) or big landslides (frane) occur but the danger is there and cannot be understimated. So it will be much more useful to prepare maps of the existing risks than inventories of the phenomena present in the area. From one side the inventory will be rapidly obsolete after an unfavourable winter, on the other an indication of the severity of the danger will impede wrong choices of the decisio-makers, certainly non specialists in dynamic geomorphology.

4 - As prof. Mensching has just shown only in the iberic peninsula areas with subdesertic characteristics exist. If we will impede an enlargement of the phenomenon of "desertification" we have to control not only these areas but also all the surrounding ones where climate is not so unfavorable but a strong risk of degradation mainly due to man already exists.

Moreover very arid regions with high climatic "infidelity" and poor parent rocks (quartzites, serpentines etc.) are widespread in the mediterranean countries. In other cases very fertile soils are present in steep mountainous slopes of the calcareous regions (Campania, Abruzzo). They are due to a thin cover of rich volcanic ashes. With the elimination of these cinders deposits by water and wind erosion the desert will appear with bare, carstic rocks outcrops.

So delicate interventions and detailed studies must be foreseen not only in the most unfavorable localities from a climatic point of view but also where the loss can be greater.

Let us now see how from the experience of this project we can advance some suggestions to better face the "desertification" problem in Europe.

1) Anthropic "desertification" occurs almost everywhere.
2) We have to control rapidly these quite different (phenomena wild urbanization, pollution, erosion).
3) Monitoring but also efficient interventions must be foreseen.
4) For these interventions many and expensive basic documents are needed.
5) The European community will have in a few month a very good new soil map in scale 1:1.000.000.
6) Documents of these kind but at different level and therefore with the adoption of a suitable scale must be prepared at national, regional and local niveau.
7) We have to stimulate a closer coordination and collaboration between the scientific community and the elected decision-makers.
8) Not more time can be lost the solution is urgent.

5 - We all, colleagues and friends, live in the old Europe and are companions in the european community where a democratic system exists and will be preserved also in the future.

The only way to impede in Europe the enlargement of desertification, the increase of soil and land degradation is to convince the people that these problems are their own problems. N.W. Hudson, a wellknown specialist in our disciplines, stated in 1980 "that although there are gaps in the available technical knowledge the more important contraints on soil conservation are political, social and economic - not necessarily in that order of importance but together much more significant than deficiencies in techniques".

This is very true in my opinion. Politicians, decision makers will provide with good laws, sufficient financing only if they will understand that a strong popular permanent pressure will not allow inattention and negligence on these problems.

C. SOME CASE STUDIES

Desertification in Northwestern Greece; the case of Kella

The mining factor in the desertification process

Desertification due to air pollution in Attica

Minimization of photosynthesis due to air pollution

The Central Huerva Valley (Zaragoza, Spain) : A part of Sahara in Europe

Soil erosion and land degradation in Southern Italy

Aspects of desertification in Sardinia - Italy

Desertification through acidification : the case of Waldsterben

Soil degradation in a North European Region

DESERTIFICATION IN NORTHWESTERN GREECE; THE CASE OF KELLA

D. VOKOU, J. DIAMANTOPOULOS, TH.A. MARDIRIS and N.S. MARGARIS
Division of Ecology, Department of Biology, School of Sciences,
University of Thessaloniki,
Thessaloniki-Greece

Summary

The mountainous area in the northwestern Greece near the Vegoritis lake
is characterized by extremely low plant cover. According to the climatic
conditions prevailing and taking into account the vegetation of adjacent
areas it seems that the type of system favoured there is that of
deciduous trees mixed with evergreen sclerophylls. Nevertheless, the
vegetation is very sparse dominated exclusively by *Juniperus oxycedrus*;
under such conditions, soil erosion is very strong. The soil bareness
is attributed to overgrazing pressure acting for long in the area.

1. DESERTIFICATION IN GREECE

The problem of desertification is quite often very serious in Greece (2).
In general two desert-like types are encountered. The "asphodel deserts" in
the areas characterized by the peculiar mediterranean-type climate (see paper
by Pandis and Margaris in the same volume) and those occurring in areas
with more or less temperate climate. The second type was never examined in
the past. Therefore, we considered it very interesting to deal with and
to present a general overview. The area of Kella in northwestern Greece,
close to the borders with Yugoslavia is such a typical case (Fig. 1)

Figure 1. The Kella desert in northwestern Greece.

2. DESERTED AND DESERTIFIED LAND DOMINATES

Figure 2 present a map of the area, covering approximately 160 km^2. To delineate the "desertified" areas a LANDSAT false color imagery was used (date 6 AUG 80). At this summer date most of the herbaceous plants in the area are dried. As the shrubs do not have a cover more than 30%, the study area can be distinguished from the surrounding forests (which reflect the infra red and thus appear on the imagery red) and from the arable lands which at this season lye fallow, reflecting more in the visible zone.

As base map was used the sheet "Ptolemais" from the 1:100000 topographic map of Greece.

The whole area presented is divided in agricultural land and natural systems, the majority of which can be characterized as shrublands and deserts. Quantifying the surface covered by each type (Table I) it was found that desert-like systems consist 74% of the area.

The population of the five communities encountered there is less than 3000 inhabitants, half that forty years ago (Figure 3). The population density there is 17 persons.km^{-2} when in whole Greece it is 66, in Macedonia (where the Kella area belongs) 55 and in the whole Prefecture of Florina (to which all communities belong) 28. It becomes apparent, therefore, that a serious population decrease proceeds considering both time and space.

The exhaustion of resources as a result of overexploitation of the

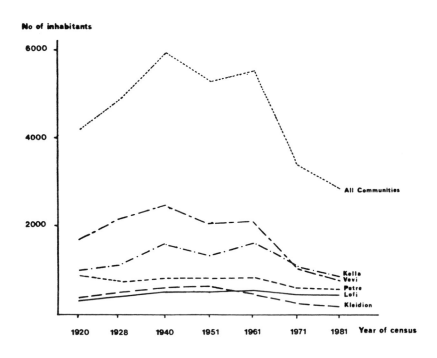

Figure 3. Population change in the five communities for the period 1920-1981.

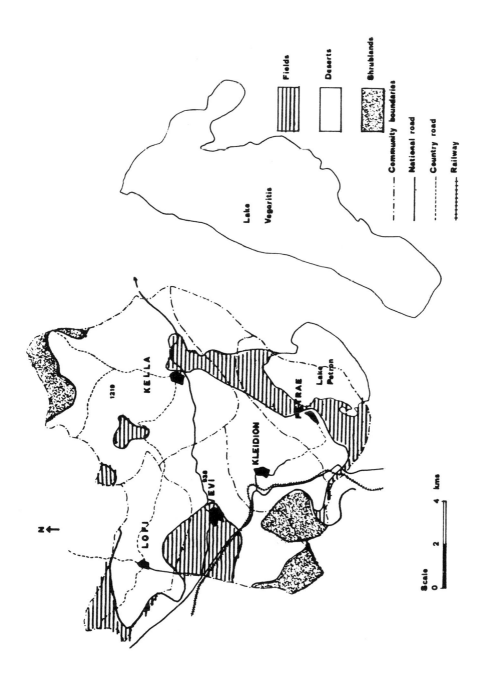

Figure 2. Map of the area examined constructed on the basis of satelite photographs and observation in situ.

Table I. Surface covered by the five communities in the area examined,
divided in cultivated land and natural systems, estimations of
the desertified land and numbers of grazing animals.

Communities	Surface (ha)	Desertified area (ha)	%	Cultivated land (ha)	Shrublands (ha)	Sheep and goats
Kella	5735	4900	86	807	4928	6620
Vevi	3123	1817	58	1317	1806	3350
Lofi	2400	1900	79	582	1818	1110
Klidion	3044	2560	87	395	2649	1400
Petre	1822	685	39	567	1255	1490
TOTAL	16124	11862	74	3668	12456	13970

natural systems must be considered as one of the main reasons which lead
to human abandonment.

3. LESS DIVERSITY-LESS STABILITY

It is undoubtedly not very easy to determine the interface between
shrubland and deserts. As an approach to give an answer we applied a
technique already used in the past(1,3)which permits us to compare the
degree of degradation in that area with that of other shrublands suffering
less from anthropogenic stress.

The type of vegetation in the area called by Turill (4) "shiblyak" is
described as a community in the place of the distructed forests; but, "many
of the characteristic species are light demanding and certainly could not
withstand the cool shade of forest".

Characteristic species are:*Paliurus spina-christi, Cotinus coggygria,
Syringa vulgaris, Prunus* sp., *Juniperus oxycedrus, Cercis siliquastrum*
whereas in the surrounding forests dominates the oak *Quercus macedonica.*
In six sites selected to represent the gradient of plant coverage we
measured in ten sampling quadrats of 100 m^2 each the number of woody species
present and the percentage of plant coverage. The relation between plant
coverage and number of woody plants can be found in Figure 4. It is clear
from this figure that the decrease of plant cover is accompanied by less
diversity. The species dominating in each of the sites sampled are shown
in Table II. The non-palatable *Juniperus oxycedrus* is found in all sites
whereas the deciduous species such as *Quercus macedonica, Pyrus amydgali-
formis, Paliurus spina-christi* are less resistant.

The degradation is very serious and the soil erosion massive. The
topsoil is continuously removed from the slopes of the hills. For example,
at the foot of a hill not overpassing 350 m the soil was more than 45 cm
deep; at a height 50-300 m it was 7-11 cm deep whereas near the top less
than 4 cm.

It is of interest to note the inefficiency of the system to support
grazing animals. Though the population is mostly pastoral only 14.000 animals
(Table I) graze in the area what corresponds to less than 1 animal per
hectare, a value representative of very poor pastures.

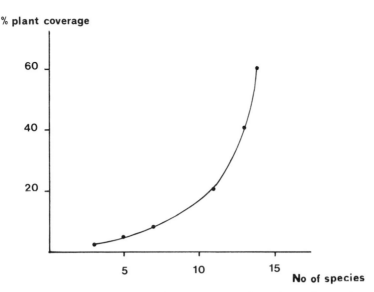

% plant coverage

No of species

Figure 4. Relationship of plant coverage and number of species present in the sites selected, along the gradient of degradation.

Table II. Dominant species in the six sites along the gradient of degradation.

Site	% coverage	Dominant species
1	60	*Juniperus oxycedrus,Quercus macedonica,Asparagus acutifolius, Paliurus spina-christi, Rosa* sp.
2	40	*Juniperus oxycedrus,Quercus macedonica,Paliurus spina-christi,Phillyrea media*
3	20	*Juniperus oxycedrus,Quercus macedonica,Pyrus amygdaliformis*
4	8	*Juniperus oxycedrus, Quercus macedonica*
5	5	*Juniperus oxycedrus*
6	2	*Juniperus oxycedrus*

4. GENERAL REMARKS

The exact causes that induced in the past the desertification process are not fully understood. However, some historical records give evidence of a clear cutting in the beginning of the century; furthermore, since the local population is mainly pastoral, overgrazing in combination with frequent fires may be considered as the bigger factor to this process.

The use of this area as pastures for freely grazing animals is unsound and costly, aggravating in the same time the situation of this degraded system. In only a few years it will not be able to support the animals that still graze there.

We believe that the first step of a rehabilitation policy should be the control of grazing. In areas with less than 100 g.m^{-2} aboveground bio-mass, grazing should stop completely. Upgradation of the system will then follow slowly but efficiently aided by the relatively high precipitation of the area.

However, the socio-economic and political infrastructure is such that even a simple theoretically measure like that of grazing control is very difficult to be implemented. The solution might be found in the acceptance of and participation to the rehabilitation project of the local population Without its collaboration there is no chance of success.

REFERENCES

1. Diamantopoulos J, 1983. Structure and distribution of the Greek phryganic ecosystems. Ph.D.Thesis (in Greek).
2. Margaris NS, 1984. Desertification in Greece. Progress in Biometeorology 3: 120-128.
3. Parsons DJ and Moldenke AR, 1975. Convergence in vegetation structure along analogous climatic gradients in California and Chile. Ecology 56 (4).
4. Turrill WB, 1929. The plant-life of the Balkan peninsula. A phytogeo-graphical study. Oxford, Clarendon Press.

THE MINING FACTOR IN THE DESERTIFICATION PROCESS

I. ROUSSIS*, D. VOKOU**, TH.A. MARDIRIS** and N.S. MARGARIS**
*National Electric Corporation, Ptolemais Lignite Centre
**Division of Ecology, Department of Biology,School of Sciences,
University of Thessaloniki, Thessaloniki-Greece

Summary

Land disturbance due to surface mining of the lignite deposits in the
Ptolemais-Amideon area (N. Greece) has resulted to desert-like conditions. The surface disturbed at present is approximately 30 km^2 and
is estimated to cover 135 km^2 by 2020 when depletion of deposits is
predicted. Though the exploitation of lignite started in 1955 no
serious efforts of rehabilitation have been undertaken up to now

1. INTRODUCTION

The loss of natural productivity of a previously fertile area, described
by the term desertification, is due rather to overexploitation of resources
or the combination of climatic factors and mismanagement of a certain area.
The desertification process may expand both in time and space depending on the mediating factors.
A localized but very intense desertification process follows the
exploitation of subterrannean deposits not accompanied by rehabilitation
procedures. Such a typical example is the Ptolemais case in Greece.

2. THE SITUATION IN THE MINES OF PTOLEMAIS-AMIDEON

In the Department of Kozani, Northern Greece (Figure 1) is the biggest
plant of electric power production in the country. It functions by utilizing
the lignite deposits which abandon in the area around the town of Ptolemais
and Amideon estimated to 1.8 bilion tons. In 1982, the total consumption
of lignite for the production of electric power was about 22 million tons
what produced $\simeq$ 10 million kwh.

Figure 1. Ptolemais-Amideon region in Northern Greece

Lignite exploitation with surface mining started in 1955. In 1982 the area given to mining, depositions after the removal of lignite and installations covers ≃ 30 km². However, exploitation of all lignite deposits available will cover at the end another 107 km², what means that at the final stage of the works, predicted to be around 2020, a total area of 135 km² will have been disturbed.

The area around Ptolemais-Amideon is seprated into 8 distinct mines (Figure 2). Excepted one, their exploitation has already started and in one of them it has already finished.

Table I. Duration of mines exploitation; present and future surface covered by each mine.

Mines	Starting date of exploitation	Predicted final date of exploitation	Surface covered after depletion of deposits(ha)	Surface of works today(ha)
Main field	1957	1983	1200	1200
Field Kardias	1970	1990	1600	900
Southern field	1979	2020	4800	400
Field Komanou	1980	1994	600	200
Field Karyochoriou	1981	2000	500	100
Amideon	1982	2018	4350	50
Diavolorema	2000	2020	500	35
TOTAL SURFACE			13550	2885

From Table I it can be seen that the mining works in the Main Field, the first given to exploitation, are completed. Therefore, it is the receptor of much of the depositions from the Kománou and Karyochoríou mines and of its own as well.

The relation of the surface mined to that given to depositions after removal of the lignite is given in Table II.

Table II. Area mined and covered by depositions after removal of the lignite in each mine.

Mines	Surface mined (ha)	External depositions (ha)	Total surface (ha)
Main field	800	400	1200
Field Kardias	1200	400	1600
Field Karyochoriou	500	–	500
Field Komanou	600	–	600
Southern field	3000	1800	4800
Diavolorema	500	–	500
Amideon	2550	1800	4350
TOTAL	9150	4400	13500

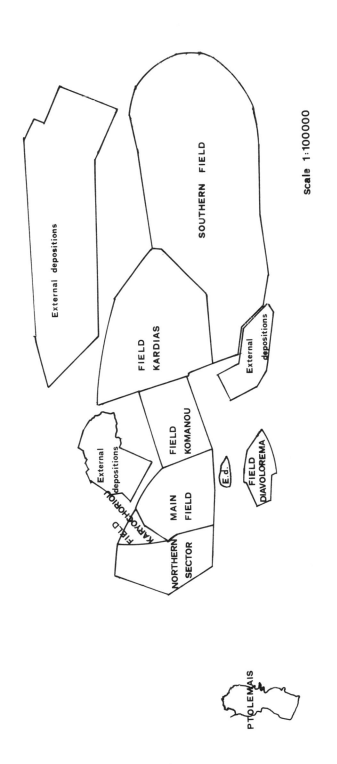

Figure 2. Map of the mines and external depositions in Ptolemais area.

By the year 2020, when it is predicted tyat all lignite deposits will have been depleted, the configuration of the land will be totally different. The area covered by each type of land (flat, inclined and depressions) is shown in Table III .

Table III. Land morphology by 2020.

	Level land (ha)	Inclined land (ha)	Depressions (ha)	Installations (ha)	Total surface (ha)
Ptolemais mines	4140	3400	740	920	9200
Amideon mines	1950	1600	350	450	4350
Total area	6090	5000	1090	1370	13550
% of the total area	45	37	8	10	100

The situation around the Ptolemais plant nowadays is that of a desert originated not by the mining activities themselves but by the short-sighted planning and indifference to the environmental disturbances. It should not be forgotten that in 600 ha of the area (Main field) the works have already finished. A serious planning should have given importance to its immediate rehabilitation.

3. PROBLEMS AND PROPOSALS

The greatest problems that should be faced in a project of environmental upgrading are the absence of any vegetational cover, the destruction of the soil structure and the extremely serious wind and water erosion resulting as a consequence of the above. Such a project should direct to the evaluation of the possible future land use, be it agricultural land, wood plantations, pastures etc. The content and availability of nutrients in the soil should be estimated. In experimental plots various crops, fruiting trees, ornamental plants should be grown in order to test the potential efficacity of the land for a productive cultivation, the yields, and the requirements for improving both the quality and quantity of the products of any sort.
Another important aspect which needs special emphasis in an integrated rehabilitation project is the social problem.
All this area was formerly agricultural land. Gradually after the initiation of the mining works the structure of the population as far as the economic activity is concerned has been altered and from purely agricultural in the past it became industrial nowadays. However, lignite deposits are to be depleted by 2020. At that time, if no measures are taken by now, local people will remain without jobs and all this area will be left both desertified and deserted. If, however, till that time the project of rehabilitation has proceeded properly and efficiently the transition

to other types of economic activity will be mild, not accompanied by great socioeconomic problems, the environment will be upgraded and the area will flourish again.

The need for such a procedure is urgent. The small group of volunteers who at present have undertaken with inadequate means to work on this direction, in spite of their hard efforts, cannot for sure solve the problem.

DESERTIFICATION DUE TO AIR POLLUTION IN ATTICA

N.S. MARGARIS, M. ARIANOUTSOU-FARAGGITAKI, S. TSELAS and L. LOUKAS

Division of Ecology, Department of Biology, School of Sciences,
University of Thessaloniki
Thessaloniki-Greece

Summary

In this work some productivity characteristics as well as the regeneration capability of two typical mediterranean-type Attican ecosystems
are considered in relation to their air-pollution load. It is found
that both the primary production and the regeneration of the most
polluted area (Korydalos) is strongly eliminated, and this finding is
taken as an obvious signal of degradation which is gradually directing
the system to a desert situation, due to air-pollution.

1. INTRODUCTION

Air pollution is known to have detrimental effects on plants, usually
at the productivity level (1,2,6,8). The perturbation happening is also
affecting systems' stability. Recently, in a project dealing with the action
of air-pollutants on Attican natural ecosystems (5), we found that in the
more stressed areas plant species diversity and productivity are strongly
eliminated, while some of the typical plants of these systems appear visual
symptoms of perturbations (see Psaras et al., in the same volume).

This paper discusses in more detail the effects of air pollution on
some characteristics of the primary production of two experimental sites,
selected among a net of stations (5) as the most representative ones.
Special emphasis is given on the systems' capability to reestablish themselves through regeneration.

2. EXPERIMENTAL SITES

The selection of the study sites was based on previous data which
concerned their productivity characteristics (5) as well as on our macro-
scopical observations. These sites are Korydalos and Varkiza. The former
is found on a slope of Egaleo mountain facing Piraeus and the latter is
Varkiza, located on a southern slope of mountain Hymettus, east of Athens.

3. EVALUATION OF THE ORIGIN OF THE STRESS; ENVIRONMENTAL OR ANTHROPOGENIC?

Petrall (7) has tried to compare sociological systems with natural ones.
For this reason, he used the percent coverage of each plant species in a
plant community as a criterion of its importance value. Analysing many
phytosociological tables, he constructed summation curves, where the
relative importance of a plant species is accumulated from the least to the
most important ones. Using summation curves constructed according to this
technique it is possible to distinguish between systems found under
natural stress and those subjected to anthropogenic ones.

Figure 1 is revealing summation curves constructed for the Korydalos
and Varkiza areas using as a parameter the plant relative coverage.

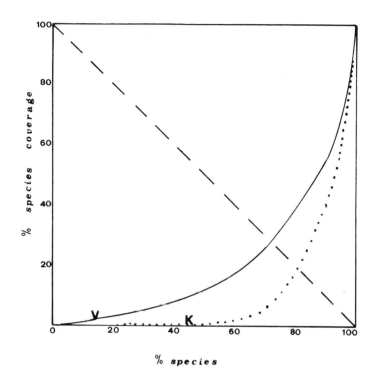

Figure 1. Petrall's summation curves for the two ecosystems studied.
        K: Korydalos, V: Varkiza.

        Although the results are relative (3), the following remarks can be
drawn:
1. It seems that both ecosystems tested are under environmental stress and
   this is something normal if we consider that these systems are mediterranean ones and they are facing the problem of water shortage.
2. From the shape of the curves we can deduce that Korydalos area is under a
   severe anthropogenic stress, since the infection point of the curve
   almost drops to the abscissa. On the other hand, Varkiza reflects the
   existence of a typical environmental stress, normally occurring in mediterranean-type ecosystems.

4. PRODUCTIVITY CHARACTERISTICS

        As it is stated in the introduction, air pollution has a direct effect
on productivity level. Table I contains data dealing with productivity
characteristics of the two selected studying areas. The information drawn
from these data are strong evidence of degradation of the Korydalos system.
For example, the total above-ground plant biomass is almost 40% lower in
Korydalos, comparing to that of Varkiza, while the same is true also for
the relative portion of the green biomass. The low percentage of the green
biomass, means of course minimization of a photosynthetic tissues, which
make the ecosystem active and productive. Furthermore, considering that
tissues appear serious structural damages (see Psaras et al., in the same
volume), the result is more severe.

Table I. Productivity characteristics of the two ecosystems studied.

	Korydalos	Varkiza
1. Total above-ground biomass $(g.m^{-2})$	152.0	245.0
2. Plant cover (%)	50.0	44.0
3. Ratio of 1/2	3.0	5.6
4. Green parts (%)	27.7	44.3
5. Leaf Area Index $(cm^2.m^{-2})$	3078.0	9254.0

5. SYSTEMS' REJUVENATION

The next step of our effort was to find whether the ecosystems are affected from the air pollution up to the level of their regeneration capabilities. For this reason we estimated the age distribution patterns in both studying areas, using as a testing plant a typical woody one, usually dominant in these ecosystems. The plant is the subshrub *Euphorbia acanthothamnos* and the method followed was the measurement of its canopy diameter and its mathematical relation with age (4).

Figure 2 makes clear that a serious perturbation is going on in Korydalos ecosystem, since the age classes 0-5 years is strongly eliminated, while on the other hand this is not valid in the Varkiza case. where a real age pyramid occurs. This fact means that the Korydalos ecosystem is loosing its vigour and it is gradually transformed to a senescent system with no young individuals, finally lacking rejuvenation.

Another very interesting point is that the age classes 5-10 years in Korydalos area are of the same order with those of 0-5 years in Varkiza and this actually suports the idea that at least 5 years ago air pollution started affecting Korydalos site. It is really peculiar, but the trigger of all political crises and people's environmental awareness connecting with air pollution coincide timely with this finding.

6. SOME GENERAL REMARKS

Without doubt our findings could be characterized as preliminary, but they do support the idea of an evercoming desertification in some of the heavier air polluted natural Attican ecosystems both on the productivity level as well as on their reproduction mechanisms. It seems that either the seeds cannot germinate or the young seedlings are not able to survive in this changed environment.

Apart of the completness or not of these data, and provided that before them we knew nothing, we consider them as very interesting and significant, since they consist the first steps in a direct action for the future management, which in nowadays is strongly needed.

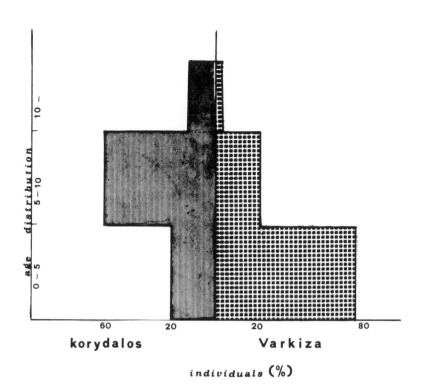

Figure 2. Age distribution of *Euphorbia acanthothamnos* Individuals. Classes
0-5, 5-10 and 10- years.

REFERENCES

1. Kozlowski T.T., 1980. Bioscience 30:88.
2. Legge AH, 1980. In: Miller PR (ed.), Effects of Air Pollution on Mediterranean and Temperate Forest Ecosystems. USDA/Forest Service, Pasific
   Southwest Forest and Range Exp. Stn, Berkeley, California.
3. Lieth H, 1975. In: H. Lieth and RH. Whittaker (eds.), Primary Productivity of the Biosphere.Springer, Berlin
4. Margaris NS, 1976. J. Biogeography 3:249.
5. Margaris NS, Arianoutsou M, Tselas S, Loukas L and Th. Petanidou, 1983.
   Tech. Rep. EEC (Contract No B 6612/9).
6. Miller PR and McBride JR, 1975. Effects of air pollutants on forest. In:
   Mudd JB and Kozlowski TT (eds.), Responses of plants to air pollutants.
   Academic Press, New York.
7. Petrall P, 1972. In: H. Lieth (ed.) Papers on Productivity and Succession
   in Ecosystems. EDF Biome, Memmo, Report p. 19.
8. Winner WE and Bewley JD, 1978. Oecologia 35:221.

MINIMIZATION OF PHOTOSYNTHESIS DUE TO AIR POLLUTION

G. PSARAS*,M. ARIANOUTSOU-FARAGGITAKI** and N.S. MARGARIS**
* Institute of General Botany,Department of Biology, University of Athens-
**Division of Ecology, Department of Biology, University of Thessaloniki-
Greece

Summary

In this work the leaf structure of *Phlomis fruticosa* and *Urginea mari-tima* from a strongly perturbated, by means of air pollution, phryganic ecosystem in Attica (Greece) was studied in comparison to the leaf structure of the same plants from a similar phryganic ecosystem free of air pollutants. The first, light microscope detected, detrimental effects of air pollutants refer to chloroplasts. The latter are fewer and smaller in the leaves of the affected plants. A second effect of air pollution on the *P. fruticosa* leaves is the shape of the chloro-plast containing mesophyll cells, which appear to be more irregular than in normal plants. Besides, leaves of *U. maritima* exhibit colorless areas, that is areas without chloroplasts, where the mesophyll cell layers have been practically reduced. In other words the leaf of *U. maritima* shows abundant colorless constricted areas in its blade. It is reasonable to presume that the above effects on chloroplasts leads to the minimization of photosynthesis. Therefore, our results justify the low primary production of the most air polluted ecosystems as well as the lower biomass estimations in these, available in the literature.

1. INTRODUCTION

It seems that the basic effect of air pollution to the plants is direct-ed towards the minimization of their productivity (1,2,4,7,8,9).
Morphological, macroscopically visible symptoms are the ultimate results of changes in the plants due to air pollution. Chlorosis of leaf tissue and breakdown of chlorophylls and other plant pigments are some of these symptoms (6).
During the last year our group started working on the effects of air pollution on Attican natural ecosystems. Our first results showed a serious decrease of plant biomass and elimination of species diversity (3). At the same time, the general appearance of the dominant plants occurring in the most polluted study areas showed an obvious perturbation and loss of vigour.
Having in mind the above, we selected two typical mediterranean plants, one geophyte (*Urginea maritima*) and a subshrub (*Phlomis fruticosa*) in order to find out the exact action of air pollution on leaf structure.

2. MATERIALS AND METHODS

The selection of the two study sites was based on the information avail-able to us, concerning the biomass yield (3) as well as on our empirical estimation of industrial emissions. These sites are Korydalos and Varkiza. Korydalos found on a slope of Egaleo mountain facing Piraeus, is a smog-covered area, with low biomass, therefore it is considered to be a strongly perturbated area. Varkiza, located on a southern slope of Hymettus mountain, east of Athens, is considered a non-air-polluted area, yielding to normal levels of biomass.

Leaves from *Phlomis fruticosa* and *Urginea maritima* individuals were collected keeping in mind that the plants had to be of similar phenology and their leaves had to be of the same age (Figures 1,6 and 11,18 asterisk). Micrographs presented here are coming from the same areas of the leaf blade correspondingly. Therefore, histological differences do not reflect the examination of different sites of the leaves.

Plant material was fixed in 6% glutaraldehyde in 0.025M phosphate buffer at pH 7 at room temperature for 3 hours. Postfixation was carried out in 1% OsO_4 in the same buffer. The plant material was washed in buffer and dehydrated in a gradient of aceton solution. Then it was placed in propylene oxide for 30 minutes and embedded in Durcupan ACM (Fluka). The sections (1-3 µm thick) were stained in 1% toluidine blue O in 1% borax solution according to Pickett-Heaps (5) and observed with a Zeiss light microscope.

3. RESULTS AND DISCUSSION

The leaves of *Phlomis fruticosa* (Figure 1) as they are seen in cross section are typically bifacial (Figure 2). Abundant multicellular hairs cover both the upper and the lower epidermis (Figure 2). Upper epidermis consists of one layer of flattened cells with moderately thick external wall. Lower epidermis comprises smaller thin-walled cells of irregular shape (Figure 3). Stomata occur in the lower surface of the leaf only and they are characteristically raised above the rest epidermal cells surface (Figure 3). One layer of long palisade and three to four layers of small and nearly iso-diametric spongy parenchyma cells form the ground tissue of the leaves (Figure 3). Abundant large and lens-shaped chloroplasts are present in both the palisade (Figure 4, arrows) and the spongy parenchyma cells (Figure 5, arrows).

The leaves of *P. fruticosa* from the smog suffering system exhibit several green-yellow areas (Figure 6). As it is seen in cross sections, the leaf structure remains unchanged in respect of the number of cell layers of each tissue, i.e. of the epidermis (cf. Figure 7 with Figure 2), of the palisade (cf. Figure 8 with Figure 3) and spongy parenchyma (cf. Figure 10 with Figure 5). On the other hand disturbance is evident firstly in the shape of all kinds of cells of dermal and ground tissues (cf. Figures 8,9, 10 with 3,4,5, respectively) and secondly in the number, the shape and size of chloroplasts. It is evident that the number of chloroplasts is strongly eliminated in palisade (cf. Figure 9 with Figure 4) and in spongy mesophyll cells (cf. Figure 10 with Figure 5). Moreover the chloroplasts of the air-pollution affected plants are smaller and seem to be more flattened than the ones of non-perturbated plants (cf. Figures 9,10 with 4,5 respectively).

The leaves of *Urginea maritima* (Figure 11) are almost unifacial (Figure 12). The thickness of the blade is unique (Figure 12), except of the central vein and the marginal zone; the about 10 cell layered mesophyll comprises two or three layers of cells containing chloroplasts at each side (Figure 12). Abaxial chlorenchyma shows larger intercellular spaces and more irregular shaped cells (Figure 14) than the adaxial one (Figure 13).

Leaves of most plants of the air polluted area exhibit abundant pale areas on their surfaces (Figure 18). As seen in cross sectioned leaves, these areas consist of the upper and lower epidermal layers while mesophyll layers have been practically reduced from 10 (Figure 12) to 2-3 cell layers (Figure 15, open arrow-heads).

In distal areas from the narrow zone (Figure 15, upper area in the left) chloroplasts of chlorenchyma cells (Figure 16) seem to be the same in number and shape with the ones of undisturbed plants (Figure 16, cf. with Figures 12,14). Closer to the restriction (Figure 15, upper area in the middle) chloroplasts seem to be fewer and smaller (Figure 17, cf. with Figure 16 and Figures 13,14) whole parenchyma cells of the narrow areas do not contain chloroplasts (Figure 15).

The above data show clearly that the two plants, selected as indices from the perturbated area, appear considerable elimination of photosynthesis, because of the pollution-affected leaf structure. This fact undoubtedly leads the system to low productivity and degradation.

REFERENCES

1. Kozlowski TT, 1980. Bioscience 30:88.
2. Legge AH, 1980. In: PR Miller (ed.), Effects of Air Pollutants on Mediterranean and Temperate Forest Ecosystems. USDA/Forest Service, Pasific Southwest Forest and Range Exp. Stn., Berkeley, California.
3. Margaris NS, Arianoutsou-Faraggitaki M, Tselas S, Loukas L and Petanidou D, 1983. Tech. Rep. 2, EEC (Contract No B 6612/9.
4. Miller PR and McBride JR, 1975. Effects of air pollutants on forest. In: Mudd JB and Kozlowski TT (eds.), Responses of plants to air pollutants. Academic Press, New York.
5. Pickett-Heaps JD, 1969. J. Cell Sci., 4:397.
6. Posthumus AC, 1982. In: Steubing L and Jäger H-J (eds.), Monitoring of air pollutants by plants. Methods and Problems. Dr. W. Junk Publishers, The Hague/Boston/London.
7. Winner WE and Bewley JD, 1978a. Oecologia 33:311.
8. Winner WE and Bewley JD, 1978b. Oecologia 35:221.
9. Winner WE, 1981. In: Margaris NS and Mooney HA (eds.), Components of productivity of Mediterranean-climate regions. Basic and Applied Aspects. Dr. W. Junk Publishers, The Hague/Boston/London.

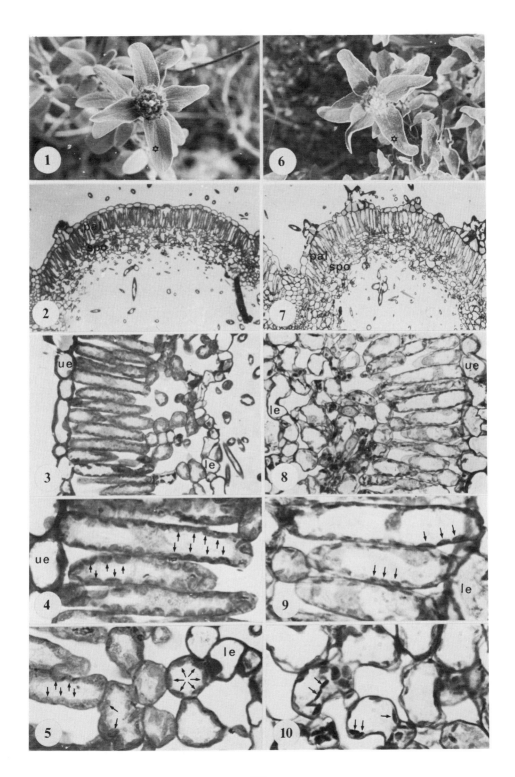

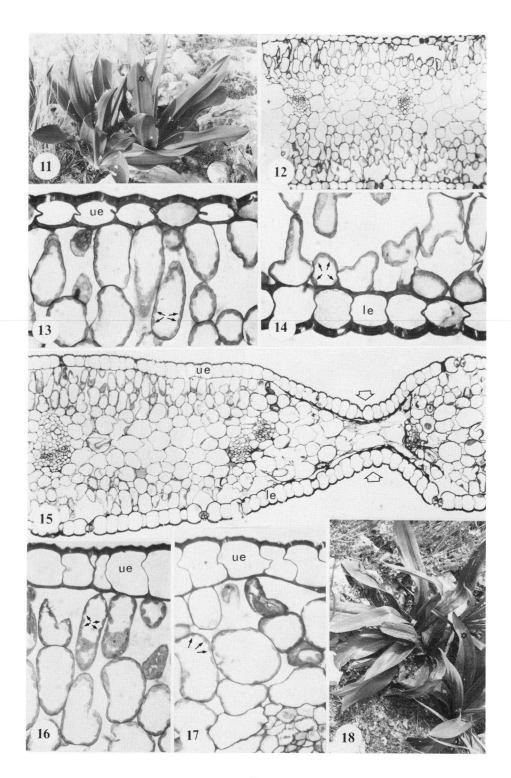

LEGENDS

Figures 1-10. *Phlomis fruticosa*

Figures 1-5.   Plant material from an area free from air pollutants (Varkiza).

Figures 6-10. Plant material from an area where the presence of smog is evident (Korydalos).

Figures 1,6.   The two last pairs of leaves. Samples were taken always from the second pair of leaves (asterisk). Notice that perturbated leaves (Figure 6) show several light areas (Both X2/3).

Figures 2,7.   Light micrographs of cross sectioned leaves (X70).

Figures 3,8.   Cross sections of leaves·note the abundance of chloroplasts in Figure 3 in comparison to those in Figure 8 (Both X300).

Figures 4,9.   Palisade parenchyma cells from cross sectioned leaves; chloroplasts are indicated by arrows (X700).

Figures 5,10. Spongy parenchyma cells from cross sectioned leaves; arrows indicate chloroplasts (X700).

Figures 11-18. *Urginea maritima*

Figures 11-14. Plants from an undisturbed area (Varkiza).

Figures 15-18. Plants suffering from smog (Korydalos).

Figures 11,18. The whole plants (Both X1/10). The leaves of the perturbated plant (Figure 18) show abundant pale areas.

Figures 12-14. Cross sections of a leaf (Figure 12, X70). A portion of the upper epidermis with chlorenchyma is magnified in Figure 13 (X300). A portion of the lower epidermis with some chlorenchyma is magnified in Figure 14 (X300). Chloroplasts are indicated by arrows.

Figures 15-17. Cross section of a leaf. Narrow areas (open arrow heads) with fewer mesophyll cells without chloroplasts are present at the pale spots of the leaves (Figure 15, X70). Chloroplasts seem to be not affected in shape and number in distal areas from the narrowing (Figure 16, X300) while they seem to be fewer and smaller in proximal regions (Figure 17, X300).

THE CENTRAL HUERVA VALLEY (ZARAGOZA, SPAIN):A PART OF SAHARA IN EUROPE

J. DIAMANTOPOULOS
Division of Ecology, Department of Biology, School of Sciences,
University of Thessaloniki, Thessaloniki-Greece

Summary

The combination of mediterranean climate with limestone and gypsum
rocks creates a very arid environment in the study area. Vegetation
consists of dwarfscrubs where dominate *Genista scorpius*, *Thymus vulgaris*,
Rosmarinus officinalis (Romero) and the grass *Brachypodium ramosum* of a
very low nutritive value. On the "gypsum rock" grow the above mentioned
plants as well as specialized dwarfscrubs such as *Salvia lavanduli-*
folia and *Linum suffruticosum*. Main land uses are: cultivation of wheat
on the flat areas, vines and almonds on the smooth slopes and
grazing on the steeper slopes. Erosion is accelerated. Productive
potential is very low and it cannot be increased significantly with
the usual improvements (anti-erosion action, irrigation). The possible
effects of alternative land uses in a context of carefull environmental management are discussed.

1. INTRODUCTION

The Ebro valley is long ago considered, after Almeria and Murcia, an
outstanding example of a desertified area under an arid climate (6).
Although in Northern Europe the term desertification might be considered
equivalent to "environmental degradation" (see Grove in the same volume),
for the Southern Europe which is closer to the true desert, the term seems
more appropriate since the desert is not always a degraded environment even
if its productive potential is low.
Man acting on the land and its vegetative cover, in order to increase
his income, seems to be the main factor in the development of "deserts" in
places where they ought not to be; but there are other factors also interrac-
ting like climate, tectonics and geology which accentuate or attenuate
man's impact.
Ebro valley offers an excellent example of such a land where happened
some mistakes of management especially after 1950 but in which nature plays
a very active role in the desertification process because by the geologic
substratum and its geomorphology determines land use and the techniques of
management.

2. SITE DESCRIPTION

The site is situated slightly north of the contact between the Iberian
System and the Ebro Basin along the central Huerva Valley, between the
villages of Tosos and Muel, in the Province of Zaragoza. The topographic
height range between 400 and 700 m (7).

Figure 1. Location map.

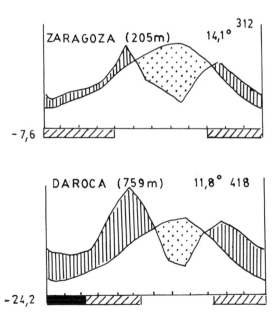

Figure 2. Climadiagrams of the two nearest stations (data from Walter and Lieth, 1960 (8)).

3. GEOMORPHOLOGY AND LITHOLOGY

The main geomorphological units of the area, important from a vegetation and agriculture point of view are: (a) structurally controlled plateau; (b) structural (anticlinal, synclinal) ridges; (c) badlands both active and slightly active; (d) fluvial terraces and (e) infilled valley floors ((1). Table I presents these main geomorphological units along with their corresponding lithology (important for structural influence as well as for vegetation). Land use and vegetation are directly influenced by geomorphological and lithological properties of these units as it will be shown in the following paragraphs. Since each geomorphological unit encompasses, vegetation and soil characteristics of each the term landscape unit will be used also in this text.

Table I. Main geomorphological units, their lithology and the corresponding land uses.

Geomorphological unit	Main corresponding lithology	Land use
Structurally controlled plateau	Limestone, marl and gypsum (Miocene)	Arable crops
Structural (anticlinal, synclinal) ridges	Limestone (Jurassic)	Grazing, beekeeping
Badlands active and slightly active	Gypsum, marl, clay and siltstone (Miocene)	Grazing on gypsum, vines, almonds on siltstone
Fluvial terraces	Fluvial deposits	Horticulture
Infilled valley floors	Infilled valley and depression deposits	Arable crops

4. CLIMATE

Climate is typically mediterranean, with rainfalls during autumn and spring. From the meteorological station of Cariñena which lies on al altitude of 590 m and has as coordinates 44°20′22′′N - 02°28′00′′E we have a mean annual temperature of 12.8°C and mean annual rainfall 594 mm.

The course of mean monthly temperature and precipitation is shown in Figure 2 (8).

From the above mentioned data we conclude that the total amount of rainfall is satisfactory for the needs of agriculture; moreover, its concentration in fall and spring offers a temporal distribution pattern better than the one shown by other mediterranean climatic regimes where rainfall is concentrated mainly in winter.

In spite of the above mentioned favorable conditions from the point of view of rainfall the area appears desert-like although other mediterranean areas with lower annual precipitation show a denser and higher vegetation cover.

Studies of the physical and chemical properties of the soils in the area are missing but we can assume that the small Water Holding Capacity

Table II. Clustering of the most frequent species in relevees over the
whole area. Numbers represent cover in a scale from 0 to 9.

```
Genista scorpius          1 7 1 1 1 2 2 2 2 3 2 2 1 2 2 2 2 2 2 1 1 1 1 7 2 2
Thymus vulgaris         2   1 2               2 1 1   2     2 2 1 1   4 7 3 1
Brachypodium ramosum    1 1   2                   3 1 7 4 4     2 3 4
Cistus clusii               1       3         1
Pinus halepensis            9                 1               6
Quercus coccifera       1             2               1           1
Juniperus oxycedrus                                 3         1 2 1
Eryngium campestre            1     2     1       1 1         1   2 1
Phlomis lychnitis           1         1                     1     1 1
Poterium sanguisorba    2             1 1     1       2 2       1   2
Hieracium pilosella     2 1       1         4   1       2 1       1 1 3
Carlina corymbosa                     5                     1   2 2
Plantago holostema          1 1                               1   1
Retama retama         1                       1
Helichryssum italicum             7           1           1 1
Carex hallerana              1       1 1                 2
Helianthemum cinereum         2             2
Koeleria vallesiana           1       1             3
Sedum sediformis        2   1 1   1             1     1 1       1 1
Helianthemun marifolium       2       2           1
Medicago sp.                                    1           2
Koeleria sp.              1               1       2           4
Festuca sp.                                      2 2     2     2
Juniperus phoenicea     1             1                         1
Digitalis obscura       1 1         1                 2     1
Bupleurum fruticosum
Argyrolobium zanonii                                1           1
Rhamnus alaternus                        1
Helianthemum pillosum
Rosmarinus officinalis  2               1               1
Orobanche sp.
Lavandula latifolia              2           2 7             1
Fumana ericoides                     1 1                       1
Helianthemum apeninum
Salvia lavandulifolia                                           1
Atractylis humilis             2                               1
Asphodelus fistulosus
Bromus madritensis
Artemisia herba alba
Teucrium capitatum
Linum suffruticosum                                           1
Artemisia barrelieri
Stipa offneri
Santolina chamaecyparyssus
Santolina sp.
Lithodora fruticosa                  1
Rhamnus ludovici                        1
Rhamnus lycioides
Euphorbia serrata
```

```
7 2 1 5 4 3 3 3 2 2 2 1 1 4 4 4 2 1 2 2 1 1 1 1 1 1 5       3 3 3 3 3 3 3 2 2 2 2
2 1 2 7 3 5 2 3 2 3 4 2 1 2 7 3 1 2 1 2 3 2 5 1 2 2   2 3 5 3   6 3 3   3 2 2 2 4   2 2
2         4 1 1 4 1 3 4     4   2 3 5 2 2 7 1 1 5 1 2 2 1     5 2 2 4   2 5 4 5 2 2 5 2
  1                               3               2   4                               2 2
2                 2             6       7   4                                         7
3     4           2           1       3       7   1 1   1                         3       4
                                      1           1               5                   2       3
      1 2 1             1 1       1 1 1       1   1   2 1     1 1 1 1   1     1 1 1 1 1   2
       ·1     1 1       1       2 2         1     2   1       1 2         1   1       2
    2     2       1         1       1 1   1                 1       1       1               1
    1     2             1                                 2             2 1       1
          2                     2 1                           2           2 1         1 1
                                                             1 1   -         1   2
          1   2                     2   7                     1         2       2 2 1         2
      1 2           1         2               2 3           1         1     2 2 1
2 1 1       1   1                         1 1                     1
          1                       2                     1
                                  2         1 1                 1               2
                                  1                           2               2
1   1       2 2 1 2   2             2   1 2       2 2       1 2       1 1           1 1
2   1       1             2   1 1   1 2 1     2 2 5 1       3         1 2   1 1           1 1
          1       3                             3                     3
    2       3       8 5         2                                             3       2
                            1   2       i   1                             3     2
      1     1           1         1   2       2       1 1   1   1
        3                 1     2   1   2       2     1 1 1           1   2
          1             1         1     1     1                   1 2
                        2             1                     3
2         1   1                         2 1                 1
7                 1         7 1 6 2 2 1   4 5   1       1 4   1           4 8 2 3
        1                           1 1     1                                     1
        3                 3 2   2 1   1     1   2     1   4   1 3       2   1
                          1       1 1                   1       1   1
                                                                     1
                    1         2       1       2       1       2 3
                        3     2       2     1       2   1
                          1       1     1       1 1       1   2
                          1     2                   1
            1   2               1                 1         2
      1       1         1           1 1   2 1       1       1
      1         2 1 1       1 1   5         1
                2 1           4   1                   1
      1   1 2 1         1                 4
              1             2 1 2         2
      1                   1   1       1
        1       1 1         1           1   1 1
```

– 180 –

as well as the presence of high concentrations of gypsum seems to be the main reason for this desert like appearance of the landscape.

5. VEGETATION

The climax vegetation of the area according UNESCO (6) is characterized by *Quercus coccifera* and *Rhamnus lycioides*; according Montserrat(4), it is characterized by *Quercus ilex* ssp *rotundifolia* and other dwarfscrubs.

In general these statements are true; but if we examine the study area in larger scale a different pattern arises. *Quercus ilex* is the dominant plant in the area but it is found only over quartzites and slates and not in the study area which has other parent rocks (2).

Rhamnus lycioides is found only in the limestone and gypsum areas while *Quercus coccifera* is present all over the area but it does not dominate in the landscape.

Data collected in the field show that species dominating in the area, both in biomass and physiognomy are: *Genista scorpius, Thymus vulgaris* and the grass *Brachypodium ramosum*. The presence of these species is attributed to human influence (6).

Table II shows the three groups of plant species of the area.

The first group is composed of plant species without any particular preference for a given substratum (in the area) or species colonizing abandoned fields. Plant species of the group are found between the species of the following two groups.

The second group consists mainly of *Rosmarinus officinalis, Lavandula latifolia* and *Fumana ericoides* as well as some of the species of the first group. It is found mostly over the Jurassique limestone areas.

The third group consists of species which tolereate the high concentrations of gypsum and are found almost exclusively on such kind of rock. Vegetation cover in these areas is very low (under 30%) and also the height of plants is low (under 30 cm). Since the area is overgrazed it is difficult to estimate to what degree the low plant cover and height is due to overgrazing and to what extent to the salty environment.

6. LAND USE

Flat areas are used for wheat cultivation and after harvest for stubble grazing.

The Jurassique limestone areas are mainly used for beekeeping since the soil layer is too thin for cultivation or cultivation is forbidden in order to protect the Mesalocha dam from filling up with eroded sediments.

The "gypsum badlands" is the area which best deserves the name of desert since as it is stated earlier, vegetation cover is very thin. They are used mainly as grazing lands and locally as wheat fields where the terrain is flat.

The most valuable products of the area (vines and almonds) are produced on hilly terrain having siltstone as parent rock, where slopes are not very steep.

7. DISCUSSION

Since none of the above mentioned landuses is very profitable,planners and local decision makers are seeking alternative ones.

Irrigation during the drought period (summer) seems not to be a convenient way even though water is available from the dams of Mazalocha and Tosos. Difficulties arise from the hilly nature of the terrain which creates transportation problems and also from the unknown comportment of the soils to irrigation, in contrast with what was considered as a solution twenty years before (5).

Reforestation with *Pinus halepensis* is already done in some places; particularly successful examples are forests near the village of Vilanueva de Huerva and Tosos. Success lies not only in the fact that trees are well encroached and soil erosion is stopped but also in the fact that other plants are coming into the forest, like *Quercus ilex*, forming intermediate vegetation strata and creating in this way a more stable ecosystem from the ecological point of view.

It has to be said that reforestation is done only in the siltstone areas; there are no data of other attempts on areas with different geological substratum but it seems that are almost unsuitable for such kind of land use.

Reforestation may not make the inhabitants of the area immediately wealthier but it is the most reasonable way of land-use in the area, which besides stopping erosion and upgrading the natural environment, makes a wild and renewable source which could be used when it will be needed (3).

The other landscape units present peculiar characteristics so they should be conserved in their present day status.

The Jurassique limestone unit with the Romerales vegetation is used for beekeeping activities and this seems the best fit land-use although there are no available economic data for this kind of activity.

Gypsum badlands are used mainly as grazing lands but the poverty of the vegetation as forage is reflected in the small number of sheeps grazing in the pastures of the community of Mesalocha (1890) compared to 3900 in the nearby village of Vilanueva de Huerva which has its grazing lands mainly in Miocene siltstone (Data of the year 1978).

The usual range improvements, such as seeding of new species varieties, fertilizer application to increase foliage, may be of doubtful success since they will be applied on a special soil rich in gupsum. On the other hand they may be harmful to the special flora (and maybe fauna) which is so well adapted in this environment. So it is suggested that the area should be considered under protection.

In this way the adaptations of the plants to this environment, but also different management techniques on a special and difficult environment could be studied better.

The plateau unit, as it is said before, is entirely devoted to wheat cultivation. Since its top layer is one of the gypsum layers it presents the same problems though in a lesser degree, as the other gypsum badlands but its flatness makes it very suitable for different activities. A first step in anyway is to plant trees in the borders in order to protect from the strong wind. As far as it concerns alternative more remunerating uses such as glasshouses, a more detailed study is needed, since the soil is not well developed and contains large quantities of gypsum.

REFERENCES

1. Boyer L, 1981. Generalisation in semi-detailed geomorphological mapping. ITC Journal.
2. Bujarrabal EP, Guiral J, 1980. Montes de Tosos (Zaragoza).Aspectos Ecologicos (Internal report of ICONA).
3. Margaris NS, 1981. Maquis biomass for energy: Costs and benefits. In: Margaris NS and Mooney HA (eds.) Components of Productivity in Mediterranean climate regions. Dr. W. Junk Publishers, The Hague, pp 238-242.
4. Montserat-Recoder P, 1966. Vegetation de la cuenca del Ebro. P. Cent. pir. Biol.exp. 1(5), JACA.

5. Roquero de Laburu C, 1964. L'utilisation du sol dans la région semi-
   aride de l'Espagne. In: Land-use in semi arid Mediterranean climates,
   p. 75-80, Arid Zone Research, Unesco, Paris.
6. UNESCO, 1977. Map of the world distribution of arid regions MAB Technical
   Notes 7.
7. Van Zuidam RA, 1976. Geomorphological Development of the Zaragoza Region,
   Spain. Processes and Landforms related to climatic changes in a large
   Mediterranean river basin. Dr. Thesis, Univ. of Utrecht-ITC, Enschede,
   The Netherlands.
8. Walter H and and Lieth H, 1960. Klimadiagram Weltatlas, Fischer Verlag,
   Jena.

Data for this report were collected during an ITC-Vegetation Department
Fieldwork in the area, during the academic year 1980-1981 under the super-
vision of Prof. I.S. Zonneveld and Dr. H. van Gils.

SOIL EROSION AND LAND DEGRADATION IN SOUTHERN ITALY

H.M. Rendell
Geography Laboratory, University of Sussex, UK

Summary

Severe soil erosion and landsliding affect 17% of the land area of
Italy. The semi-arid zones of the south of Italy and the islands of
Sicily and Sardinia are particularly prone to erosion by virtue of a
combination of climatic factors and high relative relief. Such areas
are also prone to hazards such as forest fires. Soil conservation
projects have achieved some success, but in some areas problems of
erosion are being exacerbated by changes in agricultural land use.

1. INTRODUCTION

Soil erosion and landsliding pose a mjor threat to settlements, infrastructure and agricultural productivity in Southern Italy. The severe soil
erosion and landsliding in this area can be regarded simply as the
inevitable results of a combination of high relative relief, erodible soils
and climate. It could be argued that recent changes in land use have merely
served to intensify processes of degradation that were already occurring.
Man has exercised a strong influence over the vegetation cover of Southern
Italy for thousands of years through deforestation, fire and overgrazing
(1). But recent changes in land use practice involving increased mechanization and increased field sizes appear to have had an impact, in some
areas of the South, that is qualitatively different as well as more severe
than that of the traditional forms of human intervention.

In this paper, the problems of erosion experienced in the semi-arid
areas of the Italian Peninsula and the islands of Sicily and Sardinia will
be discussed in general terms. The physical environment of these areas, and
recent trends in land use will be considered before moving to a discussion
of the particular problems of the area of highly erodible clay hillslopes
in the Region of Basilicata.

2. PHYSICAL BACKGROUND

In Italy, mountainous and hilly terrains predominate with only 23% of
the land area classified as plains. In the Regions of the South, the
relative extent of plains is, in some cases, much smaller. In Basilicata
and Calabria, for example, only 8-9% of the land area is classified as
plains. Many areas of the South have high relative relief and associated
problems of slope instability. The area as a whole is tectonically active.
Geologically, Southern Italy exhibits a great diversity of rock types from
igneous and metamorphic to limestones, sandstones, shales and clays.

It is possible to identify areas potentially at risk from soil
erosion on the basis of purely climatic criteria. For the purposes of land
use capability, two zones have been identified within Southern Italy (2).
First, the sub-humid zones of Lazio, Abruzzo, Campania and the Tyrrhenian
slopes of Calabria, with mean annual moisture deficits of 400mm and
second, the semi-arid zones of Puglia, Basilicata (Province of Matera),
the Ionian slopes of Calabria, Sicily and Sardinia with moisture deficits

of over 600mm (see Fig. 1). The rainfall distribution in these zones is highly variable both year to year and spatially. Data for the period 1975-1979 give annual rainfall totals ranging from 407mm to 980mm for the semi-arid zones, with between 9 and 15% of the total falling in the summer months. The variability in annual rainfall amounts is exemplified by data from Pisticci in Basilicata, where the minimum and maximum values for a 46-yr record are 401mm and 1601mm and where the standard deviation about the mean annual value of 713mm is 217mm. It is apparent that in areas of high relative relief and erodible soils, that may be subject to intense rainfall, the maintenance of vegetation cover is particularly important in preventing or reducing the threat of soil erosion. It is no surprise that erosion is greatest during the rainstorms that mark the end of the summer's drought in the South of Italy and the islands, when ground cover

Fig. 1: Southern Italy

will tend to be at a minimum. The removal of vegetation cover, particularly in arable areas, prior to these early autumn rains exacerbates the
erosion problem.

3. LAND USE AND LAND USE CHANGES 1970-1980

Patterns of land use in the semi-arid zones reflect various strategies
for combatting seasonal, as well as more long-term, water shortage. Three
different land use strategies may be identified; first, dry farming techniques, maximising the use of rainfall for the extensive cultivation of
cereals, particularly wheat; second, the small-scale intensive cultivation
of vines, olives and horticultural produce in inland/upland areas making
use of natural springs, wells and other local sources of water supply; and
third, the large-scale intensive cultivation of plains areas mostly for
horticultural produce, such as tomatoes and strawberries, aided by large-
scale irrigation works. Land use for forestry is important in the South.
Between 1950 and 1970 over 100,000-ha of land were afforested in
Basilicata and Calabria. In the period 1970-1980 the woodland areas in
the South increased as follows: 8.1% (13,812-ha) in Basilicata, 6.06%
(24,243-ha) in Calabria, 13% (24,352-ha) in Sicily and 20.5% (67,086-ha)
in Sardinia. It is worth noting that between 0.6 and 2.5% of the forested
areas in various regions of southern Italy are destroyed by fire each
year. In 1980, of the 5660 forest fires reported for Italy, 1588 (c.28%)
occurred in the Regions of Basilicata, Calabria, Sicily and Sardinia and
affected a total area of 19,580-ha. Although a relatively small percentage of the total forest stock is affected by fire in any one year, the
local impact in terms of the onset of erosion and landsliding may be
severe. Large areas of Calabria and Sardinia were devastated by forest
fires during the summer of 1983.

Agricultural production in Italy as a whole increased by over 20% in
the period 1970-1980, but there was an overall decline in the agricultural
workforce during this period until, by 1980, the percentage of the male
population engaged in agriculture had fallen to 13%. Although the percentage of the workforce engaged in agriculture has always been higher in the
South than in the North, even in Basilicata the percentage fell from 40%
in 1973 to 33% in 1980. Agriculture in the South appears to have become
polarized between the highly mechanised extensive cultivation of grain and
the intensive irrigation-based cultivation of fruit and vegetables on the
coastal plains. The cultivation of small upland plots of land has
declined, with many plots being abandoned. Throughout the South the upland
areas have suffered a net out migration of population to the coastal
plains and elsewhere in Italy. Emigration is, of course, the classic
response to the problems of unemployment and low incomes experienced in the
upland agricultural communities. The plight of many of these declining
settlements was highlighted in the aftermath of the Southern Italian
Earthquake of November 1980.

On the basis of regional statistics, one important, if subtle, trend
in land use in the South can be identified. Between 1970 and 1980, the
area devoted to the cultivation of wheat in Basilicata and Puglia
increased. There was a strong trend towards the cultivation of 'hard'
wheat (gran duro) and the areas brought into production within Basilicata
were formerly poor pastureland. An additional point to note is that 62%
of the area of wheat production in Basilicata is hill-land. The
increases in wheat hectarage in Basilicata and Puglia run counter to
trends for the rest of Italy and for other Regions in the South, and, in
the case of Basilicata the trend is particularly interesting given the
relatively low productivity per hectare.

In general terms land use changes in the South have involved the
increased mechanisation of agriculture and the imposition of monocultures.
Although some of the land use changes outlined above might be thought to
have a potentially beneficial impact on processes of land degradation,
this is not always the case. The widespread planting of Pinus radiata in
Sardinia has, for example, had a disastrous effect on the species
diversity of indigenous flora. Also, afforestation is not the panacea for
erosion problems that it might first appear, as will be illustrated in the
case of the intensely eroded clay hillslopes of Basilicata.

4. SOIL EROSION AND SOIL EROSION RATES

Damage resulting from flooding, landsliding and soil erosion within
Italy is estimated to cost between $1.05 and $1.35 billion each year (3).
In the case of settlements, roads, bridges, farmland, watercourses and
irrigations systems the damage may be obvious, but equally important,
although more insidious, is the damage to the economic life of reservoirs
caused by rapid siltation. According to Alexander (3) 17% (or 50,000 kmsq)
of the land area of Italy is affected by severe soil erosion and/or landsliding (dissesti idrologici) and 1% (or 2,500 kmsq) is classified as
'badland' (calanchi). In the South of Italy the scale of the erosion
problem is immense. Data for sediment transport in several of the major
southern rivers give for example sediment loads equivalent to 1159 tonnes/
kmsq/yr (Bradano before construction of S. Giuliana dam), 2458 tonnes/
kmsq/yr (Sinni) and 1003 tonnes/kmsq/yr (Crati) that contrast strongly with
values for the Tevere (Tiber) at Rome of 377 tonnes/kmsq/yr and for the
Arno near Pisa of 250 tonnes/kmsq/yr (4,5). Erosion rates on individual
hillslopes are, in certain circumstances, far higher than the erosion
rates, based on sediment loads, might suggest. Even in Tuscany, Panicucci
(6) calculated erosion rates equivalent to 1347 tonnes/kmsq/yr for
unvegetated south facing Pliocene clay hillslopes while data for unvegetated Plio-Pleistocene clay hillslopes in Basilicata indicate that erosion
rates may be of the order of 28000 tonnes/kmsq/yr in extreme cases (7,8).

5. EROSION RATES, EROSION PROCESSES AND THEIR CONTROL

Particular combinations of climate, high relative relief and soil
types make soil erosion and landsliding inevitable in the hilly and
mountainous areas of the South. In the areas where the soil parent
material is a relatively hard rock, soil cover may be thin and erosion
will be limited by the availability of material. No such limitation
exists in the clay areas of the South where evidence of erosion can be
masked by ploughing and where, in the case of these poor regosols (9)
erosion is limited by the efficacy of erosion processes.

In areas subject to severe erosion and landsliding, the understanding
and measurement of processes and process interactions is difficult.
Processes of erosion operate and interact at a variety of temporal and
spatial scales. In any one area the hillslopes may be subject to erosion
by rainsplash, surface flow (overland flow/sheetflow), rill flow, concentrated subsurface flow ('piping') and combinations of any or all of these
processes with shallow massmovement. Small channels may be blocked by
landslides from hillslopes or as a result of bank collapse. In this
context the areal erosion rates calculated on the basis of sediment loads
of streams, such as those quoted above, have little meaning in terms of
the qualitative impact of erosion actually occurring in the catchments.
Even at the scale of a particular hillslope, the same amount of material
may be moved downslope by shallow landsliding or by a combination of sheet

flow and rill flow, but the impact on the slope, and the possible
remedial actions,will be very different. The quantification of the
impact of erosion and landsliding even within small catchments is difficult and estimates of 'damage cost' or 'area lost to agriculture' are
merely rather crude surrogates. Research work on erosion and erosional
landforms within Italy has tended to focus on either the description and
classification of particular landforms or terrains (10) or on the monitoring of small catchments or experimental hillslope plots (11). Recent
research funded by the Italian Government, through the Consiglio Nazionale
della Ricerche, has focussed on the problems of marginal lands with a view
to stopping or reversing the various processes of land degradation.
Emphasis is placed upon 'the defence of the soil', which is seen as an
essential first step in any programme to improve agricultural productivity.
Capital investment in the development of irrigation systems, the stabilisation of water courses and the 'reorganisation' (sistemazione) of hillslopes and small catchments has been the dominant feature of State intervention in Southern Italy, excluding funds for industrial development. The
use of physical measures to control erosion presupposes an understanding
of the erosion processes involved and this has evidently not always been
the case. In some areas conflicts of interest arise between agricultural
land use and conservation practice. The intensely eroded clay hillslopes
in the Region of Basilicata provide one example of an area in which such
conflicts of interest are currently occurring.

6. THE CASE OF THE PLIO-PLEISTOCENE CLAY AREAS OF BASILICATA

Outcrops of inorganic Plio-Pleistocene clays of marine origin cover
an area of 195,700-ha within Basilicata and account for 35% of the area of

Fig. 2: Calanchi near Pisticci, Basilicata

Fig. 3: Biancane near Pisticci, Basilicata

the Province of Matera. These clays are up to 500m thick and rapid incision during the Late Pleistocene has resulted in the development of valley
sideslopes in excess of 30 degrees in part of the middle Basento valley.

6.1 Erosional landforms

Extensive areas of badland amounting to 32,125-ha are found within
the zones of moderate to high relative relief within the Bradano, Basento,
Cavone and Agri valleys (12). Two distinct types of badland landform may
be identified in the clay areas. The term 'calanchi' is applied to the
type of knife-edged erosion features (shown in Fig. 2) that occur on
slopes of 40-60 degrees. 'Biancane' are erosional landforms that form
typically hummocky terrain (see Fig. 3), an individual biancana may be
up to 4m high. Areas of calanchi are found in zones of Miocene and Pliocene 'scaley-clays' (argille scagliose), but calanchi are most spectacu-
larly developed in the Plio-Pleistocene clay areas. Alexander (8) quotes
a figure for drainage density of 1000 km/kmsq for these badland areas; and
the rates of erosion of such slopes are very high (7,8). The clay areas
also show extensive development of pipes and pseudokarst features. The
susceptibility of these clays to the development of natural pipes, or
tunnels, as a result of the sectional enlargement of cracks, joints and
fissures within the clay rock by percolating rainwater was noticed as
early as 1745 when Antonini (13) mentions this peculiar property of the
clay.

6.2 Erosion processes

The reason for the high potential erodibility of the Plio-Pleistocene clays in this part of Italy lies in the fact that the clays are naturally 'dispersive' (14). In the presence of water, dispersive clays pass spontaneously into suspension. Since dispersive clays cannot be identified by conventional soil tests, such as the determination of Atterberg Limits or by triaxial testing, a series of special tests has been devised by soil scientists in the USA and Australia (15). The dominant clay mineral in these Plio-Pleistocene clays is sodium montmorillonite although both kaolinite and illite are also present. Sodium dominates the exchange complex thereby facilitating spontaneous dispersion in the presence of water, and the development of the natural pipes or tunnels on the clay hillslopes. The behaviour of these clays is complicated by their swelling and slaking characteristics but their dispersive properties can be suppressed by the addition of calcium hydroxide. Work on samples of the Plio-Pleistocene clay from grassland and woodland areas indicates that dispersion is also suppressed in the presence of very small amounts of humic materials. Unfortunately, once ploughed, and even when small amounts of humus are present, the clay hillslopes are highly susceptible to rill erosion as a result of the relatively impermeable nature of the clay and its susceptibility to the development of an impervious surface crust within minutes of the onset of a rainstorm. Such storms, particularly those occurring in the early autumn after the summer drought, tend to be particularly erosive in the clay area. Rainfall intensities of 1-2mm/min are not

Fig. 4: Contour ploughed and afforested slope, Grottole, Basilicata

Fig. 5: Bulldozing calanchi near Craco, Basilicata

unusual and in some cases these storms last for over an hour. The most intense 24 hr record is for Pisticci, in the Basento valley, with 314-6mm in November 1959. Work on the erosion within small ephemeral flow catchments within the clay area has indicated that substantial incision and changes in channel form can occur within a few years (16) with substantial amounts of material being pulsed through the tributary channel systems and into the major river channel.

6.3 Erosion control

Attempts to combat erosion in the clay areas of Basilicata include the afforestation of slopes, contour ploughing and the construction of check dams. Afforestation is one of the commonest responses to erosion problems on these clay hillslopes, but it may be an inappropriate one. The attempt at afforestation of the hillslopes shown in Fig.4 has evidently achieved only limited success. Some trees simply will not grow in these poor regosols. A better approach would appear to be to revegetate the slope with scrub vegetation which will at least ensure reasonable ground cover and may provide a first step towards eventual afforestation. The function of the check dams is to impose an artificial base level on the tributary channel system rather than simply to trap sediment. In one example in the Basento valley, the check dam, having filled with sediment, then suffered erosion by channelled subsurface flow by virtue of the dispersive nature of the material such that the structure was effectively undermined by pipes thereby failing to serve either of its functions. In extreme cases small areas of calanchi (badland) have simply been bulldozed into new slope forms ready for growing wheat (Fig. 5). Examples of this type of landscape remodelling are relatively rare within the Plio-Pleistocene clay area and the extent to which they represent a successful response to the calanchi problem are still difficult to assess.

6.4 Recent changes in land use

The increasing area of pastureland on the clay hillslopes being ploughed up for grain production has undoubtedly increased local erosion rates. Slopes with angles of up to 30 degrees have been ploughed with the aid of caterpillar-tracked vehicles. The susceptibility of the clay hillslopes to rill erosion once they have been ploughed has already been mentioned. The intense rainfalls in August 1982 and August 1983 within the clay area, resulted in flashy runoff from these ploughed hillslopes with soil being stripped off the slope surfaces down to plough level, inundating many roads with mud. Such changes in land use run counter to all the effort that has been put into soil conservation in these areas, and may have serious consequences for reservoir siltation.

7. CONCLUSIONS

Over 50,000 kmsq of land in Italy is subject to severe soil erosion and landsliding and considerable damage is done each year to settlements, roads, bridges and irrigation systems. Although these problems are not confined to Southern Italy (the largest area of calanchi (badlands) within Italy is in the North in the Region of Emilia-Romagna) the concentrated settlements and poor infrastructure in the upland areas of the South are particularly vulnerable. Although the public works carried out on soil conservation have achieved some success a vast problem still remains. Recent changes in land use have exacerbated erosion problems particularly in the clay areas of Basilicata.

REFERENCES

1. PIGNATTI, S. (1983). Human impact in the vegetation of the Mediterranean Basin. In Holzner, W., Werger, M.J.A. and Ikusima, I. (Eds.) Man's impact on vegetation 151-161
2. VICINELLI, D. (1963). I grandi complessi irrigui nel piano di sviluppo del Mezzogiorno. In Cassa per il Mezzogiorno Dodici Anni 1950-1962 Vol. II parte i: Attività di Bonifica
3. ALEXANDER, D.E. (1982a). In Italy 'creeping disaster' speeds up. Geotimes (November 1982) 17-20
4. PANNICUCCI, M. (1972). Richerche orientative sui fenomeni erosivi nei terreni argillosi. Annali Ist. Sperimetale Studio e Difesa Suolo 3 131-146
5. CAVAZZA, S. (1962). Sulla erodibilità dei terreni di alcuni bacini calabro-lucani. In Morandini, G. (Ed.) l'erosione del suolo in Italia II: aspetti geografici 177-196
6. PINNA, M. (1962). Lo studio del trasporto solido dei corsi d'acqua nel quadro delle ricerche dell'erosione del suolo. In Morandini, G. (Ed.) L'erosione del suolo in Italia II: aspetti geografici 41-60
7. RENDELL, H.M. (1982). Clay hillslope erosion rates in the Basento valley, S. Italy. Geografiska Annaler 64A (3-4) 141-147

8. ALEXANDER, D.E. (1982b). Difference between 'calanchi' and 'biancane' badlands in Italy. In Bryan, R. and Yair, A. (Eds.) Badland geomorphology and piping 71-87
9. MANCINI, F. (1974). Cenni illustrativi della geologia, geomorfologia e pedologia della Basilicata Giornale Botanico Italiano 108 (5) 203-209
10. IPPOLITO, F. and COTECCHIA, V. (1954). Le frane ed i dissesti nelle medie valli dell'Agri e del Sinni in Basilicata. Geotecnica 2 1-23
11. VITTORINI, S. (1971). The effects of soil erosion in an experimental station in the Pliocene clay of the val d'Era (Tuscany) and its influence on the evolution of slopes. Geographica Debrecina 10 71-81
12. RADINA, B. (1964). Contributo alla conoscenza del dissesto idrogeologico del versante Jonico-Lucano. Bollettino Società Naturalisti Napoli 73 211-265
13. ANTONINI, G. (1745). La Lucania discorsi. 612
14. SHERARD, J.L., DUNNIGAN, L.P. and DECKER, R.S. (1976). Identification and nature of dispersive soils. J. Geotech. Engng. Div., Proc. ASCE 102 (GT4) 287-301
15. SHERARD, J.L. and DECKER, R.S. (Eds.) (1977). Dispersive clays, related piping and erosion in geotechnical projects. ASTM STP 623
16. RENDELL, H.M. and ALEXANDER, D.E. (1979). Note on some spatial and temporal variations in ephemeral channel form. Bulletin, Geological Society of America 90 761-772

ASPECTS OF DESERTIFICATION IN SARDINIA - ITALY.

ANGELO ARU

DEPARTMENT OF EARTH SCIENCE UNIVERSITY OF CAGLIARI

1. Preface

This problem is being treated at the European level for the first time and is of fundamental importance to the future of the countries of the old continent.

By desertification here we not only mean environmental degradation but also the phenomenon of destruction of the primary resources and of the impossibility of reconstructing them not even on a long term.

This concept, at least for the time being, is to be referred to the Mediterranean area, where the most evident phenomenon of degradation is given by the erosion of the soil by action of the run-off waters, determined by various causes. Undoubtedly, desertification may be the resul of phenomena that changes from country to country.

The aim of this preliminary note is to present the phenomenology and the most evident results of the phenomenon of degradation in Sardinia, an island almost in the centre of the Mediterranean.

2. The Causes

Among the causes that determine soil erosion, one of the most important is given by fires of the bushes and forests. This phenomenon is repeated almost regularly in time, because in most cases it represents an intervention aimed at freeing the surface from bushes that cover the soil and at increasing the area of the pastures. Following each fire, there is an erosion of the superficial horizons of the soil by run-off waters until the outcropping of the parent material.

According to recent studies, fires moreover determine an accumu lation of water-repelling substances in the first centimeters of the soil, thus causing an increase in the run-off coefficient, and consequently an increase in solid transport. Among the forms of human intervention aimed at the speculative utilization of the land, we have "productive forestation" with fast-growing species for industry. Particularly the mountain areas undergo extensive ploughing without any thought for the protection of the soil, the control of the waters, and without considering the different hydrological characteristics of the soil, or the rate of rainfall both in relation to the exigencies of the species and to the risks of erosion.

On the plains, the phenomenon of desertification is correlated with pollution, especially from heavy metals, and with the loss of land through urbanization. The importance of these phenomena must be correlated with the low availability of the soil and the great difficulty in regenerating it rapidly.

The problems connected with the erosion of the soil by desctruction of the bushes and forests have existed since Roman times, while those connected with pollution are extremely recent.

3. The Landscapes of Sardinia

The island of Sardinia is characterised by sundry forms of land scapes related to the lithological, geormorphological, pedological and vegetational characteristics.

These aspects may be summarised as follows.

3.1. The Landscapes of the Paleozoic Metamorphic Rocks

They are represented by arenaceous, silty formations and by sha les. For all these, steep forms correspond to the more acid rocks and less steep forms to shales and more clayey shales. At the base of the mountain formations, which only in some cases are over 1000 metres, we find more or less deep colluvial formations, distinguishable from the decrease in slope.

The soils originating from these materials are Rankers,Cambisols and Luvisols (the last being almost exclusively on the colluvium),but the stretches with outcropping rock are wide.

The outcropping of the rock takes place particularly with the more acid rocks and in those areas where erosion has been more intensive. The most developed soils (Cambisols and Luvisols) tend to disappear by erosion, which has been very fast in the last decades.

3.2. The Landscapes of the Limestones and Dolomitic Rocks

The landscape of the karstic areas constitutes one of the most suggestive and fascinating environments of the island and of the Mediterranean. Unfortunately the areas with soils and vegetation (Quercetum Ilicis) which have been preseved are very few. In fact following the desctruction of the forests due especially to repeated fires, there has been an intensive erosion of the soils in very short time.

At present it is estimated that the percentage of these areas covered by soils and vegetation does not exceed 2/3%, while in some places there is clear evidence (soils and plants) showing that an old and much more developed vegetation once existed.

On the limestone material, the soils are represented by Chromic Luvisols and Orthic Luvisols, whose evolution is extremely slow.

The recovery of desertified lands is in this case extremely slow even over very long times.

Considering that these lands are interesting from the scientific and landscape point of view, a very intensive and careful intervention is necessary for their preservation in that they are a testimony of landscapes which can no longer be reconstructed.

3.3. The Landscapes of Volcanic Rocks

Volcanic rocks vary from basic rocks to acids, to which diffe-

rent forms and soils correspond. At present many areas present exten
ded outcroppings of rock and soils of minimum thickness.

The soils on the basic volcanic rocks (basalts) are represented
by Rankers, Eutric Cambisols and Andosols, while the Dystric Cambisols, especially, are diffuse on the more acid rocks.

However on these formations there are stretches more or less wi
dc with soils which are well-developed but which are totally or par-
tially destroyed by erosion following fire.

The high fertility of the soils, especially those derived from
the basalts, has allowed, since immemorial times, a considerable demographic pressure connected with a great quantity of cattle. These
socio-economical aspects too, have contributed towards an increase of
degradation, particularly in the area where the acid volcanic rocks
outcrop.

3.4. The Formations of Limestones and Marls of the Miocene

The landscape on these formations is always undulated with more
steep forms only in presence of the more compact limestones.

The soils are represented by Regosols, Cambisols and Vertisols,
normally placed in characteristic "catena". On these formations the
pedogenesis is rather fast and therefore we cannot speak of real desertification but rather of intensive degradation.

Moreover, since we are dealing with soils of high class land ca
pability, it is necessary to proceed rapidly to intensive programmes
of defence and conservation.

3.5. The Pleistocenic and Olocenic Formation

In Sardinia, these formations occupy 15% of the surface with re
sepct to the whole island.

The Medium and Inferior Pleistocene is largely more diffuse with
respect to the Superior, while the Olocenic and present floods are of

very modest extension.

On these formations, the desertification phenomena are rather li mited but in certain cases very evident in presence of zones with a high pollution rate from heavy metals, mining and industrial dischar- ges. These are generally irreversible processes due to the nature of the polluting materials.

Conclusions

The problem of desertification is therefore of great interest, especially in the islands and in the south of Italy.

The concept of desertification, as explained above, principally regards the destruction of the soil-resources and the impossibility of reconstructing it in a short time.

The process of desertification is connected principally to utili zation (e.g. fires), the most important being productive reforesta- tion and other interventions that have caused pollution.

DESERTIFICATION THROUGH ACIDIFICATION: THE CASE OF WALDSTERBEN

W. BACH

Center for Applied Climatology and Environmental Studies, Department
of Geography, University of Münster, Robert-Koch-Str. 26,
D-4400 Münster, West-Germany

Summary

It is safe to say that theories which relate Waldsterben (forest die-
back) to any one single cause fail to explain this complex phenomenon.
There is a host of contributing factors including, beside SO_2, NO_x and
acid rain, ozone, fluorine, heavy metals, viruses and pests, forest
management and climatic factors. There is a consensus among scientists
that pollutants weaken the trees' resistance so that they can become
more easy prey to other adverse influences. Beside the trees, buildings, animals and humans are affected, too, and man's very livelihood
is severely threatened when soil and water become too acidic. Acidification leads to podzolization which, in the temperate zone, results in
a steppe-like landscape with widespread abandonment, a process that
can be appropriately dubbed as desertification. Responsible for all
this is the inefficient combustion of fossil fuels and, in particular,
the tall stack policy which, instead of scrubbing the pollutants from
the gas stream at the source, allows them to spread over wide areas
thereby turning acidification into a worldwide problem. The present
type of Waldsterben, which has not been observed before, proceeds at
an unprecedented speed. To reverse the adverse trend also requires unprecedented efforts. This paper, therefore, after giving an overview
of the extent, symptoms and mechanisms of the damages, emphasizes the
active control strategies, such as more efficient and hence less fuel
use, abatement techniques at the source, and substitution of pollution-free fuels. This is supplemented by relief strategies, such as
liming, spraying, soil treatment etc. The ongoing research in this
area at our institute is briefly discussed.

1. INTRODUCTION

An observer of forests in Central Europe and elsewhere will no longer
have any difficulty in finding stands of scrawny firs and sagging spruces.
These are the visible signs of forest die-back, or Waldsterben, as it is
called in Germany. Part of this phenomenon is caused by acid rain, a term
usually used for the total atmospheric acid deposition. Its components can
be classified into three categories (41): wet deposition (rain and snow),
dry deposition (particles and gases), and special events (dews, frosts and
fogs). Acid rain stems above all from the emission of sulfur oxides (SO_x)
and nitrogen oxides (NO_x) into the atmosphere by industrial and transportation sources. In the atmosphere these substances are transformed into sulfu-
ric acid (H_2SO_4) and nitric acid (HNO_3) through oxidation and hydrolysis.
These acids and their corresponding sulfates and nitrates are transported
through and eventually removed from the atmosphere to cause adverse effects
including acidification and demineralization of soils, reduction in forest

and agricultural crops, destruction of man-made materials, degradation of
drinking water, and impairment to human health. Beside SO_2, NO_x and acid
rain, there is a host of other contributing factors to Waldsterben including
ozone, fluorine, heavy metals, parasites, forest management and climate factors. There is agreement among scientists that Waldsterben is a complex phenomenon and that it is, above all, the impact of atmospheric pollutants
which is responsible for the trees' demise. The best strategy would therefore be to reduce drastically the pollution emission at the source.

We are in the midst of this process of acidification which, as we know
from large parts of the Erzgebirge, will sooner than later turn our forests
into steppe-like landscapes with widespread abandonment and much hardship,
a process for which the term desertification is quite appropriate. To docu-
ment this ongoing destruction I start by reviewing the extent and types of
injury and by discussing the possible mechanisms of damages. I then high-
light the destabilization process in forest ecosystems caused by stress and
strain leading eventually to acidification, podzolization and desertification. This is followed by a discussion of the most promising control measures which could slow down the ongoing acidification process. Since governmental control action is woefully inadequate and may therefore not be able
to reverse the adverse trend in time, I also present relief strategies that
can possibly increase the resistance of plants or restore the conditions
which are conducive to a subsequent regrowth.

2. DOCUMENTATION OF DAMAGES

The influence of emissions in the vicinity of industrial sources on
the health of plants and people has been recognized and documented since
the 17th century first in England and Sweden and later in Germany and Austria
(9). But it was only through the tall stack policy, widely adopted during
the last two decades, that the acid rain phenomenon has turned into a world-
wide problem affecting now large areas not only in Europe (14, 26, 19), North
America (16, 30, 5), and China (22), but also in many remote areas of the
world (17), such as Bermuda (25), the North Atlantic Ocean (21) or the Amazon
rainforest (20). Table I shows clearly that in earlier years the damaged
forest areas in Europe were small and restricted to the vicinity of the industrial areas, while in the past decade or so they have increased greatly
including now many remote areas.

Table I   Development of emission damages to European forests (40)

Year	Area damaged/ha	Remarks	Sources
CSSR			
1960	40 000	Only Erzgebirge	Materna, 1960
1967	250 000-300 000	CSSR	Materna, 1969
Poland			
1961	80 000	Only Upper Silesia of	Sierpinski, 1968
1968	240 000	which 48 000 dead or dying	Sierpinski, 1968
1973	260 000		Paluch, 1973
1980	379 000		Zimny, 1980
GDR			
1960	68 500	By communication	Wentzel, 1967
1965	220 000		Ranft et al., no yr.
FRG			
1907	9 000	Smoke damage of which	Reuss, 1907
1960	50 000	31 000 in the Ruhr District	Wentzel, 1960
1982	562 000	Local inventories	BML, 1982, 1983
1983	2 545 000		(Ministry of Agric.)

The Federal Republic of Germany (FRG), with a relatively large share
of forest land, is not only one of the most severely affected countries,
but it has also conducted the most detailed damage inventories hitherto,
so that it can serve as an example of what might be in stock also for other
countries. Fig. 1 shows for 44 test areas the development of the disease
status in 556 spruce and 1675 fir trees in Baden-Württemberg over a sequence
of seasons. The rapidity with which a large precentage of these trees has
turned from healthy to sick is quite apparent. While in the fall of 1980
all spruce trees were still healthy, by spring 1983 a healthy tree could
no longer be found in either tree species.

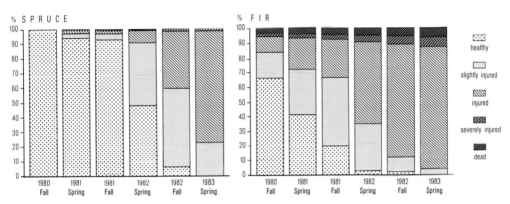

Fig. 1   Development in disease status of spruce and fir trees in Baden-
Württemberg (43)

The results from a nationwide inventory for the two years 1982 and 1983
are now also available. Although not entirely comparable, because different
methods were used, the obtained results do nevertheless indicate that injury
has become more severe and more widespread affecting now all trees in all
parts of the FRG. Table II indicates that coniferous trees, and above all
fir trees, are most severely affected, but that deciduous trees are by no
means unaffected. The overall injury rate seems to have quadrupled from 1982
to 1983. Table III shows that the injured forest area in the FRG has increased
from ca. 0.5 mill. ha in 1982 to 2.5 mill. ha in 1983. Most severely affected are the southern German lands of Baden-Württemberg with every other tree
injured as well as Bavaria and the more heavily industrialized Northrhine-
Westfalia. The damage figures in the other lands would very likely be
equally high, if the same representative sampling method, as used in these
three lands, had also been applied to the rest of the country. Finally, the
tabulation by severity of injury in Table IV shows that the individual
shares within the injury classes have increased from 1982 to 1983 - but not
equally. The share and the increase in injury class "very heavy" remained
small, because trees in this category experience a preferred cut whilst
they are still of some economic value.

3. SYMPTOMS AND TYPES OF DAMAGES

Clues for the onset of the destruction process can be obtained from
cross sections of tree trunks. Fig. 2 shows the evenly-spaced tree rings
in a healthy tree (bottom part) and the narrowing of the rings since 1960
in a damaged tree (upper part). The injury to trees occurs both above and
below ground. The main symptoms for coniferous trees above ground are the
"lametta syndrome" and the "pigment bleaching syndrome" (11). "Lametta syn-

Table II   Areal extent of injury by type of tree for 1982 and 1983 in the
            Federal Republic of Germany (8)

| Type of tree | 1982 | | | 1983 | | |
	Total forest area (Mill.ha)	Injured area (Mill. ha)	Injury of total (%)	Total forest area (Mill.ha)	Injured area (Mill. ha)	Injury of total (%)
Spruce	2.93	0.27	9	2.951	1.194	41
Pine	1.90	0.09	5	1.464	0.636	43
Fir	0.16	0.10	60	0.176	0.134	76
Beech	1.30	0.05	4	1.250	0.332	26
Oak	0.60	0.02	4	0.615	0.091	15
Other	0.40	0.03	4	0.950	0.158	17
Total/Av.	7.29	0.56	8	7.406	2.545	34

Table III   Areal extent of injury by Federal Land for 1982 and 1983 in the
            Federal Republic of Germany (8)

| Land | Injured forest area (x 10^3 ha) | | Percent of forested area | |
	1982	1983	1982	1983
Schleswig-Holstein	26	16	18	12
Lower Saxony	124	165	13	17
Northrine-Westfalia	72	295	9	35
Hesse	41	120	5	14
Palatinate	6	180	1	23
Baden-Württemberg	130	645	10	49
Bavaria	169	1115	7	46
Saarland	3	9	4	11
FRG	562	2545	8	34

Table IV   Areal extent by severity of injury for 1982 and 1983 in the Fede-
            ral Republic of Germany (8)

| Injury class | Injured forest area (Mill. ha) | | Percent of forested area | |
	1982	1983	1982	1983
Light	0.419	1.846	6	25
Heavy	0.108	0.635	1.5	8.5
Very heavy	0.035	0.064	0.5	0.9
Total	0.562	2.545	8	34

drome" means that especially older trees have shed several of their younger
needle years and that only one or two needle years hang in clusters from the
secondary shoots making the tree appear transparent. Fig. 3 shows on the
left a healthy pine branch which has five needle years and to the right a
damaged branch which is left with only two needle years forming small clusters at the outer ends of the twigs. Pigment bleaching usually starting
at the needle tips used to be restricted to urban and industrial agglomerations, but in recent years it has spread also to remote areas such as the
Lower Alpes and the Black Forest.
        The invisible damages below ground to the rhizosphere are perhaps even
more important. Fig. 4 shows on the left a cross section through a young
healthy root 3 cm off the tip, i.e. the area active in water uptake. In a
healthy root the surface is smooth, the area between central cylinder and

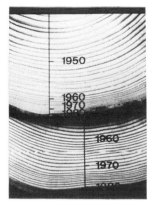

Fig. 2 Comparison of tree rings   Fig. 3 Healthy and damaged pine
        from damaged tree (above)
        with healthy tree (below) (27)

the crust is not interrupted, and the lateral roots are intact (24). The
damaged root on the right in Fig. 4 shows a brittle surface with an exfolia-
ted rhizodermis. The endodermis is in the process of disintegration. Such
a root may still be able to transfer some water upward, but an active control over the type of ions to be transported is no longer exerted. The lateral roots die off, open wounds remain, and pathogens penetrate the roots,
thereby reducing the resistance of the plant.

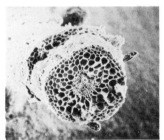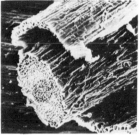

Fig. 4 Healthy (left) and damaged root (right) (24)

    Several hypotheses have been put forth on the propagation of disease
symptoms especially in coniferous species. These have been divided into
three classes (12):
    - damage of the rhizosphere system by acid rain, aluminum (Al^{3+}) toxi-
      city and depletion of minerals in the soil
    - intoxication and primary effects on leaves and needles by SO_2, NO_x, O_3,
      PAN (peroxyacetylnitrate) and other air pollutants
    - infection by as yet unknown parasites such as viroids, viruses, myco-
      plasmas or bacteria.
    No doubt more than one primary event may be responsible for the dif-
ferent disease symptoms, which may represent various stages in a "complex
disease" (13). Based on these symptoms one may distinguish two main areas
of disease in Bavaria: A north eastern region with Bayerischer Wald and
Fichtelgebirge where both atmospheric SO_2 concentration and sulfur contents

in coniferous needles are high; here the main causes of disease may be the effects of acid rain on mineral depletion and toxic SO_2 impacts on primary biochemical metabolic events. In southern Bavaria, including the northern Alpes, there is neither a high atmospheric SO_2 concentration nor a high sulfur content in needles. Moreover, the relatively low acid precipitation leads to no significant deficiency of minerals such as Mg^{2+}, Ca^{2+}, Mn^{2+}, and K^+, on these predominantly calcareous soils. Therefore, Elstner et al. (13) suspect as forcing functions of chronic intoxication fluctuations of high and low levels of photooxidants. Their studies based on the analysis of certain biochemical changes and on scanning electron microscopy seem to support the following hypothesis for the chain of events:

- photooxidants cause pigment bleaching
- the cuticles, especially the waxplugs covering the stomata are damaged
- after partial loss of both structural and physiological resistance to fungal parasites, infection by pathogens may occur
- needles invaded by fungal pathogens abscise after necrotisation
- the "lametta syndrome" and dying branches or trees are the visible manifestation of these stages of events.

4. STRESS AND STRAIN THROUGH ACID RAIN

The concept of stress and strain used in physics, biology, and geology may be applied to shed some further light on the role of acidification in the complex disease called Waldsterben. The exertion of a stress on a system results in a strain (29). Such a strain may be elastic (reversible) or plastic (irreversible), the latter meaning that the system has permanently changed some of its characteristics. A plastic strain may be invisible (latent) or visible whence it is called an injury.

Table V shows that the driving force for a forest ecosystem is climate (37). A large variability both in heat and in moisture can damage the root system, thereby reducing the elastic strain to almost all stress factors with the result that such plants fall more easily prey to plastic strain. In recent years trees have also become more susceptible to the mechanical elements of climate such as wind, snow and ice which cause increasingly wind fall and crown breaking indicating that the elastic strain may already have been critically reduced. Also the chemical climate has been dramatically changed by the increasing release of air pollutants as is evident by the more than 10fold increase in the input of acidity and chemical elements into Central European forest ecosystems. This corresponds to a reduction in the pH-value by more than one unit. The mean monthly pH-values in precipitation collected at stations in Gelsenkirchen and Bochum have decreased from pH 5.2 (1967 - 1972) to pH 4.3 (1978 - 1980) (28). The pre-industrial estimate of pH 5.1 in rain (34) can serve as a benchmark for the low mean value of pH 3.4 in intercepted rainwater recorded presently below the spruce canopy in the Solling (36). Also the deposition of nutrients such as N, S, Ca and Mg has increased to such an extent that it may account for more than half of the trees' annual nutrient intake. Acidification of soil decreases the elastic strain to many stress factors especially in and following warm and dry years. Damages to the root system in acid soils through toxic metal cations such as Al, Fe and Mn can even prevent root regeneration, thus leading to plastic strain.

Acid soils have a low resilience so that forest ecosystems on them can become easy prey to pests (Table V). Air pollution weakens the trees making them more susceptible to pest attack. Thus, the large pest outbreaks in recent years especially in Central European forests on acid soil, may be considered as precursors of Waldsterben.

Table V Stress and strain in forest ecosystems - an overview (37)

Stress factor	Causal relationship	Plastic strain
Climate		
heat climate		
warm	acidification push	root damage
cool	deacidification phase	recovery
humidity climate		
wet	O_2 deficiency in soil	root damage
dry	water deficiency in soil	root damage
mechanical climate		
no wind	-	-
storm, snow, ice	mechanical stress to roots and canopies	
chemical climate		
normal	low nutrient input	-
close to sea coast	NaCl salt damage	crown deformation
air pollution	manyfold	leaf injury
		soil acidification
		bark (cambium) injury
		root damage
		damage to decomposers: destabilization
Consumers		
pests		
viruses		diseases
bacteria		
fungi		wood rot
insects		feeding damages
man		
biomass utilization	diminishing nutrient stocks changing microclimate diminishing feeding source for decomposers	complex: destabilization
Natural loss of ecosystem elements		
death of individuals deterioration of structural units	e.g. change in microclimate	none in stable ecosystems

Forest cultivation has been practiced in Europe for the past 5000 years. This has led to soil acidification and the destruction of large forest areas resulting in a steppe-like heath and grass ecosystem of low biomass productivity. The desertification process in the past was halted by restricted forest use and timber management. It is clear that the present desertification process can only be reversed by an immediate and drastic reduction in the release of air pollutants. Finally, the natural loss of elements from the ecosystem, if not naturally replaced, must also be considered as a stress factor. For many years reforestation is no longer possible in such areas as the Erzgebirge, and apparently beech trees can no longer regenerate naturally in forests north of the Ruhr District (18, 6).

5. DESTABILIZATION OF THE FOREST ECOSYSTEM

According to Ulrich forest ecosystems pass through different phases during their development (39). First is the aggradation phase when in the soil organic matter and mobilizable nutrients are accumulated and the typical species composition and structure of the ecosystem is developed. Next follows a quasi-steady state when the forest ecosystem recycles all nutrients inside the system to keep the soil chemistry close to equilibrium. Climate stress (through large variability) and chemical stress (through large amounts of air pollution), when exceeding the buffer effect due to silicate weathering, can upset this quasi-steady state by accumulating toxic cation acids which leads to soil acidification. This, together with the other influencing factors discussed above, leads, in turn, to a destabilization of the whole ecosystem. Again two consecutive phases of destabilization can be distinguished, namely humus disintegration and podzolization. After humus loss the ecosystem can reach a new steady-state but at much lower resilience. Podzolization leads to the elimination of trees from the ecosystem, i.e. to Waldsterben and desertification. Typical examples are shown in Fig. 5 for the Fichtelgebirge (right) and the Harz (left).

Fig. 5 Dead forest followed by grass steppe, Fichtelgebirge (right) and Harz (left)

6. CONTROL STRATEGIES

There are three main control strategies which, used in combination, can bring the necessary relief on time. These are:
- energy consumption reduction through more efficient use
- emission reduction through abatement techniques, and
- energy source substitution using non-polluting technology.

6.1 Efficient energy use

Presently about 90% of our energy use is based on fossil fuels, and this is not likely to change dramatically in the future. Therefore, the more efficiently we use energy, the smaller will be the consumption of fossil fuels, and the smaller will be the emissions of SO_2, NO_x, heavy metals and a host of other hazardous substances. The potential for improving energy efficiency is very large. Methodically convincing empirical studies from about a dozen different countries show through a least-cost energy approach, which implies that people will use energy in a way that saves them money, that in the future energy needs, and hence fossil fuel use, can be expected to go down, not up. The same or even more energy services can be offered with a lower energy input simply by increasing energy productivity (31, 4).

Without requiring any technological breakthroughs much can be achieved simply by applying present technological knowledge in a rational way. For example, the overall efficiency of conventional power plants could be increased from about 36% to 85 % simply by cogeneration, i.e. by combined heat and power generation. By building decentralized neighborhood cogeneration systems the losses incurred in transporting energy and the investment needed for grid systems could be significantly reduced. With fluidized bed combustion, not only is efficiency increased by 10 %, but also the emission of SO_2 and NO_x is reduced drastically, as is shown in section 6.2. Efficient insulation is one of the most effective means of reducing fuel consumption. Sodium-vapor, quartz-halogen lamps are now available that use 75 % less current than conventional light bulbs while lasting 4 - 5 times longer. Fuel consumption of motor cars has been reduced with existing technology from the current level of 8 - 10 l/100 km to about 2 l/100 km in a VW Rabbit (Golf) type car. This together with catalytic devices, can significantly reduce the NO_x and hydrocarbon pollution.

More specifically, Table VI shows the savings in primary energy in various heating technologies as compared to a conventional oil central

Table VI   Effect of efficiency technologies on primary energy use and SO_2 emission (15)

Measure	Savings in primary energy[1] (%)	Estimated reduction in SO_2 [1] (%)
Oil heating system (base value)[2]	0	0
Improvement in engineering design[3]	5 - 20	5 - 20
Insulation, double glazing	20 - 70	20 - 70
Heat recovery, insulation, passive-solar construction	30 - 90	30 - 90
Solar collectors plus oil heating system	25 - 50	25 - 50
Heat pump (Diesel)[4]	ca. 65	ca. 65
Heat pump (gas)[4]	ca. 65	ca. 100
Heat pump (electric) plus oil heating[5]	5 - 10	Increase by 150
Heat pump (electric monovalent)[5]	Increase by 30	Increase by 300
Direct electric nighttime heat storage[5]	Increase by 300	Increase by > 300

1) Values are not accumulative
2) All percent values are related to a conventional oil heating system; emission factor 200 kg SO_2/TJ
3) E.g. electronic sensors and thermostats, thermostatic valves, electronic fuel-air mixing regulators, etc.
4) Monovalent system
5) Emission factor 6×10^{-3} kg SO_2/kWh for electric current

heating system (15). The reduced primary energy demand translates into reduced SO_2 emissions. Most effective are the gas heat pump, passive-solar construction with insulation, and insulation plus double glazing. Note, on the other hand, the significant increase in SO_2 emission, if electric heat pumps and direct electric night-time heat storage are used.

6.2. Abatement techniques

The available abatement techniques can be grouped into three categories (32, 35):

- those that reduce the sulfur content of the fuel <u>prior</u> to combustion (coal cleaning, coal gasification, desulfurization of liquid fuels),
- those that reduce SO_2 emission <u>during</u> combustion (burner technology, fluidized bed combustion), and
- those that reduce SO_2 emission <u>after</u> combustion (flue gas desulfurization).

<u>Coal cleaning</u> -- Bituminous coal contains both pyritic and organic sulfur in roughly equal proportions. Coal washing is standard practice for the removal of ash. Washing and other physical separation techniques used to remove pyrites can reduce the sulfur content of coal by 40 to 60 % by weight at a cost of $2.50 to $3.25/t (at 1975 US prices). However, these processes cannot remove the organic sulfur in the coal.

<u>Coal gasification and liquefaction</u> -- Currently available techniques can remove up to 90 % of the sulfur from the gaseous and liquid fuels. The costs of sulfur removal are likely to be low compared to the cost of gasification or liquefaction.

<u>Desulfurization of liquid fuels</u> -- There are a number of processes available. Desulfurization of gas oil achieves a sulfur removal of 90 % requiring an additional 3.5 % of energy for the removal process. Direct residual fuel oil desulfurization can reduce the sulfur level of the residue by ca. 80 % requiring 6 - 8 % additional energy of the feedstock. The degree of desulfurization that can be achieved by the indirect method of residual fuel oil desulfurization is only about 30 - 45 %; and the additional energy consumption is about 5 %. Desulfurization plants require for planning and construction a lead time of 3 to 5 years.

<u>Burner technology</u> -- Through the use of multistage burners 80 % of the SO_2 and 50 % of the NO_x can be removed as compared to emissions from conventional burners. The additional cost of combustion modifications is generally less than 1 % of the capital cost of the power plant.

<u>Fluidized bed combustion</u> -- The boiler consists of a reaction chamber in which finely ground coal is burned in suspension over a bed of moving air in the presence of fragmented limestone capable of removing SO_2. The movement of the coal particles gives a larger heat transfer thereby making a boiler of half the usual size possible and improving the energy efficiency by 10 %. Fluidized bed combustion is most suitably used in connection with smaller-scale cogeneration plants, thereby reducing the waste heat release by 80 %, the SO_2 emission by close to 95 % and, as a result of the lower combustion temperatures (800 - 900°C as compared to the usual 1600°C) the NO_x emission by about 50 %. Many experts consider this the most promising process. Additional investment costs are less than $10/kWe (in 1982 US $).

<u>Flue gas desulfurization</u> -- There are two main processes, the dry and the wet scrubbing of stack gases. In the dry process gases pass through a bed of absorbant, such as activated coal, which reacts with the SO_2 in the flue gas. When the absorbant is fully loaded with SO_2, clean gas is passed through the bed, stripping out the SO_2, which is subsequently converted to sulfur or sulfuric acid. This method can only remove some 50 % of the SO_2 in the stack gas. In the wet process the gas is washed with an alkaline solution removing up to 95 % of the SO_2 from the stack gas. The SO_2 is converted to a waste product (sludge) or to a saleable by-product (gypsum). Catalytic methods achieving NO_x removal rates of up to 80 % are practised widely in Japan. The costs of these systems amount to about 5 % of the to-

tal cost of generating electricity. Significant advantages of this method over the coal conversion route are that the total capital and operating costs are almost an order of magnitude lower, that thermal efficiencies are higher, and that utility requirements are lower (42).

6.3 Energy source substitution

The more widespread use of renewable energy resources and nuclear energy has been suggested as a substitution strategy for the reduction of SO_2, NO_x, and a host of other hazardous substances. Except for biomass burning which would produce small amounts of SO_2 and NO_x as compared to fossil fuel burning, all other solar-based renewable resources operate pollution-free.

The potential of nuclear energy to substitute fossilfuels with the purpose of reducing the acid rain is less clear and thus warrants a closer look. In the FRG nuclear energy produces about 15 % of the electricity (which itself accounts for ca. 15 % of the end-use energy) so that it approximately supplies 3 % of the end-use energy (7). In 1980 savings in the use of heating oil alone amounted to about 10 % of end-use energy. German power stations use less than 5 % of the total oil to produce electricity. More than 53 % of the oil is used to produce heat. Thus, to make a noticeable contribution, nuclear power would have to penetrate the heat market. A power plant of the Biblis type (1300 MW) would produce ca. 6.8 bill. kWh/yr or 0.84 MTCE electricity equivalent, taking a load factor of 60 %. The present German oil consumption is about 190 MTCE. Thus 10 large nuclear power plants could just replace barely 5 % of the oil, half that experienced through savings in the heating oil sector alone. The situation is not much different in other industrial nations.

With the siting and acceptance problems inherent in nuclear power plant construction, it would take at least 10 to 30 years to put on line the 30 to 40 large nuclear power plants required as a minimum to replace the existing fossil fuel plants in Germany. This, very definitely, would come too late to halt the acid rain catastrophe.

The 30 to 40 nuclear power plants would require an investment sum of at least 190 billion marks at current prices - ignoring all corrollary costs. According to the German utility industries desulfurization of the stack gases by 50 % of the current values in all existing fossil fuel power plants would require an investment of some 6 bill. marks - or just the construction costs of one single 1300 MW nuclear power plant.

Finally, the suspicion is growing that nuclear power is also a contributing factor to Waldsterben (1). Recent forest damage inventories revealed for a number of nuclear power plants (Obrigheim, Würgassen and Esenshamm in Germany as well as Bugey in France) that forest damage is significantly enhanced downwind of the main wind direction. In the vicinity of the Karlsruhe reprocessing plant it was found that Krypton (^{85}Kr) in air, and tritium (^{3}H) in water and in pine needles had increased by a factor of 5000, 40 - 160, and 9, respectively. ^{14}C, ^{85}Kr and ^{3}H are known to reduce the enzymatic power of plants to repair damages so that the more conventional pollutants (SO_2, NO_x, O_3, heavy metals etc.) can have a greater adverse impact.

7. RELIEF STRATEGIES

The fact is that the present Waldsterben proceeds at an unprecedented speed. It is feared that the current control measures which are woefully inadequate in most countries may not be effective in time. Therefore, as supplementary measures the following relief strategies are either being carried out or under consideration:

- liming of soils and water bodies

- spraying of leaves
- utilizing protective fungi
- breeding resistant species
- strengthening a plant's detoxication and reparation mechanisms.

   Liming as a buffer against high acidity has been widely advocated (38, 3), and it is applied from the air (Fig. 6) and from the ground to both soil and water in many countries (33). Preliminary results from the Spruce Liming Experiment carried out in Northrhine Westfalia show that compared to the unlimed period (1942 - 1963) yearrings obtained from trees of a limed period (1964 - 1980) yielded increased growth (3). A liming of forest soil of the order of 2500 kg/ha is usually recommended. But there are also critical voices warning of the loss of humus and nitrogen, the leaching of minerals, the polluting of drinking water and the damage to the forest ecosystem (10).

Fig. 6   Liming of forests
         and lakes (33)

   Spraying of plants as a protective measure against pests, frost and evaporative water loss is a long-established procedure (23). Currently a lignine-like substance is sprayed on trees in a large-scale field experiment in Lower Saxony to test its usefulness as a protective layer against the impact of oxidants and other pollutants.

   Many soil fungi, such as the mycorrhiza, cover the plants' roots with a protective network of filaments and provide them with additional nutrients, hormones and water. One strain, called Pisolithus arrahyzus, allows for example, fir trees to live on high levels of acidity and metals (2). There are still some severe problems to be overcome, namely the fact that fungi seem to be species-specific implying the need to identify the "right" fungus for the respective tree, and that deleterious root fungi can outcompete the benevolent ones.

   The individual resistance of any one particular tree is determined by its genetics, and, above all, it is site-specific. Breeding of a new generation of trees takes about 50 years or more so that this measure comes too late, since the dying forest needs help now and not in 50 years (10). Moreover, the breeding of specific species leads to a reduction in genetic diversity, a dangerous trend enforced by most of man's actions.

   Throughout evolution the living cell has developed a multitude of detoxication and reparation mechanisms. These include (11):

- small molecules, such as ascorbic acid, glutathione, carotine as antioxidants
- enzymes, such as DNA and RNA, nuclease and polymerase, catalase and reductase
- metallic sequences, such as the ascorbate-peroxidase-glutathione-reductase path, and
- hormone-guided shifts from katabolic to anabolic metabolic paths.

   From long-time application in human medicine it is known that glucuronegamma-lactone, an uronic acid, increases not only the body's resistance toward an insult from fungi, bacteria and viruses, but that it also helps in the body's detoxication process. It is interesting to note that both glucuronic acid and ascorbic acid are of a similar chemical structure. It is hypothesized that these substances might also be usefully applied to plants in their struggle against acid rain and related hazards. Preliminary effects are indicated in Fig. 7 showing the accomplished growth in 1983 of a Norway spruce twig. The two growth periods within a year are clearly indicated. Normally the stronger spring growth is followed by a weaker summer growth. After the spring growth the tree was watered with a solution of glucurone-gamma-lactone the effect being clearly visible in the larger and sturdier needles in the summer growth.

Fig. 8 Norway spruce clone in climate chamber (right) and identification procedure (left)

Fig. 7 Effects of glucurone-gamma-lactone on Norway spruce

    In order to find out at which doses these substances might give the best results we have just built a climate and a germination chamber to test Norway spruce clones under controlled conditions. Fig. 8 shows one of the three-year old potted spruce trees and part of the very involved identification and measurement procedures that must be carried out prior to the start of a test series. These laboratory tests are supplemented by field experiments using Norway spruce trees of the same age group.
    A number of other experiments are planned including raising the pH-value in the soil by applying sodium silicate (Na_2SiO_3), testing the chelating potential toward Al- and other metal ions of NTA, humic, citrus, wine and unsaturated fatty acids etc., and measuring ethylene as an indicator for stress and ethane as a marker for tissue damage in trees.
    In closing one must realize, however, that all of these relief actions are doomed to failure, if they are not at the same time accompanied by a drastic reduction in air pollution through stringent controls at the source and a more efficient, and hence reduced, use of fossil fuels.

REFERENCES

1. ANON. (1984). Waldsterben. Auch Atomkraft schuldig? Natur 3, 10 - 11
2. ANON. (1984). Surviving acid rain. The Economist, April 14, 92
3. ATHARI, S. and KRAMER, H. (1983). The problem of determining growth losses in Norway spruce stands caused by environmental factors. In: B. ULRICH and J. PANKRATH (eds.) Effects of Accumulation of Air Pollutants in Forest Ecosystems, 319 - 325, Reidel Publ. Co., Dordrecht.

4. BACH, W. (1982/84). Gefahr für unser Klima, C.F. Müller, Karlsruhe (English version: Our Threatened Climate, Reidel Publ. Co., Dordrecht).
5. BILONICK, R.A. and NICHOLS, D.G. (1983). Temporal variations in acid precipitation over New York State - what the 1965 - 1979 USGS data reveal, Atm. Env. 17(6), 1063 - 1072
6. BLOCK, J. (1983). Pers. communication
7. BOSSEL, H., HOECKER, K.-H. (1982). Kernkraft pro und kontra, Natur 2, 59 - 63, and Natur 3, 40 - 43.
8. BUNDESMINISTERIUM DES INNEREN (1981). 2. Immissionsschutzbericht der Bundesregierung
9. COWLING, E.B. (1982). Acid precipitation in historical perspective, Env. Sci. & Technol. 16(2), 110A - 123A
10. DEUMLING, D. und HATZFELDT, H. (1983). Kalkung, Düngung, resistente Baumzüchtungen, waldbauliche Maßnahmen - können sie den Wald retten? In: Katalyse Umweltgruppe (Hrsg.), 190 - 193, Volksblatt-Verlag, Köln
11. ELSTNER, E.F. (1983). Baumkrankheiten und Baumsterben, Naturwiss. Rundschau 36(9), 381 - 388
12. ELSTNER, E.F. und OSSWALD, W. (1984). Fichtensterben in "Reinluftgebieten": Strukturresistenzverlust, Naturwiss. Rundsch. 37(2), 52 - 61
13. ELSTNER, E.F., OSSWALD, W. and YOUNGMAN, R.J. (1984). Basic mechanisms of pigment bleaching and loss of structural resistance in spruce (Picea Abies) needles: Advances in phytomedical diagnostics, Experientia (in press)
14. FOWLER, D. et al. (1982). Rainfall acidity in northern Britain, Nature 297, 383 - 386
15. FRITSCHE, U. (1982). Sanfter Weg statt Saurem Regen - zum Zusammenhang von Energieversorgung und Schwefeldioxid-Emissionen. Öko-Magazin, Bd. 7, 57 - 93, Verlag Bonz, Fellbach
16. GALLOWAY, J.N. and WHELPDALE, D.M. (1980). An atmospheric sulfur budget for Eastern North America, Atm. Env. 14, 409 - 417
17. GALLOWAY, J.N., LIKENS, G.E., KEENE, W.C. and MILLER, J.M. (1982). The composition of precipitation in remote areas of the world, JGR 87 (11), 8771 - 8786
18. GEHRMANN, J. and ULRICH, B. (1983). Der Einfluß des Sauren Niederschlags auf die Naturverjüngung der Buche. In: Immissionsbelastungen von Waldökosystemen, 32 - 36, Sonderheft, Landesanstalt für Ökologie, Landschaftsentwicklung und Forstplanung Nordrhein-Westfalen
19. GEORGII, H.W., PERSEKE, C. and ROHBOCK, E. (1984). Deposition of acid components and heavy metals in the Federal Republic of Germany for the period 1979 - 1981, Atm. Env. 18(3), 581 - 589.
20. HAINES, B., JORDAN, C., CLARK, H. and CLARK, K. (1983). Acid rain in an Amazon rainforest, Tellus 35B, 77 - 80
21. HARRISS, R.C. et al. (1984). Atmospheric transport of pollutants from North America to the North Atlantic Ocean, Nature 308, 722 -724
22. HARTE, J. (1983). An investigation of acid precipitation in Qinghai Province, China, Atm. Env. 17(2), 403 - 408
23. HEROLD, D. (1983). Eine Tarnkappe für Pflanzen, Natur 11, 20 - 23
24. HÜTTERMANN, A. (1983). Frühdiagnose von Immissionsschäden im Wurzelbereich von Waldbäumen. In: Immissionsbelastungen von Waldökosystemen, 26 - 31, Sonderheft, Landesanstalt für Ökologie, Landschaftsentwicklung und Forstplanung Nordrhein-Westfalen.
25. JICKELLS, T., KNAP, A., CHURCH, T., GALLOWAY, J. and MILLER, J. (1982). Acid rain on Bermuda, Nature 297, 55 - 57
26. KALLEND, A.S., MARCH, A.R.W., PICKLES, J.H. and PROCTOR, M.V. (1983). Acidity of rain in Europe, Atm. Env. 17(1), 127 - 137
27. KATALYSE UMWELTGRUPPE (Hrsg.) (1983). Das Waldsterben. Ursachen, Folgen, Gegenmaßnahmen. VolksBlatt-Verlag, Köln.

28. KUTTLER, W. (1982). Belastung für den Boden, Umweltmagazin 10, 56 - 61
29. LEVITT, J. (1980). Responses of plants to environmental stresses, II. Water, radiation, salt and other stresses. Academic Press, New York
30. LIKENS, G.E. and BUTLER, T.J. (1981). Recent acidification of precipitation in North America, Atm. Env. 15(7), 1103 - 1109
31. LOVINS, A.B., LOVINS, L.H., KRAUSE, F. and BACH, W. (1981/83). Least-cost energy: Solving the CO_2 problem, Brick House, Andover (German version: Wirtschaftlichster Energieeinsatz: Lösung des CO_2 Problems, Alternative Konzepte 42, C.F. Müller, Karlsruhe).
32. PERSSON, G.A. (1976). Control of sulfur dioxide emissions in Europe, Ambio 5 (5-6), 249 - 252
33. SCHWEDISCHES LANDWIRTSCHAFTSMINISTERIUM (1983). Die Versauerung, Staatl. Amt für Umweltschutz, Solna
34. SCHWELA, D. (1983). Vergleich der nassen Deposition von Luftverunreinigungen in den Jahren um 1870 mit heutigen Belastungswerten, Staub 43, 135 - 139
35. SWEDISH MINISTRY OF AGRICULTURE (1982). The 1982 Stockholm Conference on Acidification of the Environment, June 21 - 30, 1982, Stockholm
36. ULRICH, B., MAYER, R. and KHANNA, P.K. (1979). Deposition von Luftverunreinigungen und ihre Auswirkungen in Waldökosystemen im Solling, Schriften Forstl. Fak. Univ. Göttingen 58, 291
37. ULRICH, B. (1983a). A concept of forest ecosystem stability and of acid deposition as driving force for destabilization. In: B. ULRICH and J. PANKRATH (eds.) Effects of Accumulation of Air Pollutants in Forest Ecosystems, 1 - 29, Reidel Publ. Co., Dordrecht.
38. ULRICH, B. (1983b). Gefahren für das Ökosystem durch Saure Niederschläge. In: Immissionsbelastungen von Waldökosystemen, 9 - 25, Sonderheft, Landesanstalt für Ökologie, Landschaftsentwicklung und Forstplanung Nordrhein-Westfalen
39. ULRICH, B. (1984). Effects of air pollution on forest ecosystems and waters - the principles demonstrated at a case study in Central Europe, Atm. Env. 18(3), 621 - 628
40. WENTZEL, K.F. (1982). Das Ausmaß der Waldschäden - ihre ökologische und landeskulturelle Bedeutung in Zentral-Europa. In: Waldschäden durch Immissionen, 7 - 25, Informationstagung Gottlieb Duttweiler Institut, Zürich
41. WISNIEWSKI, J. and KINSMAN, J.D. (1982). An overview of acid rain monitoring acitvities in North America, Bull. Amer. Met. Soc. 63(6), 598 - 618
42. YAN, T.Y. (1984). Fueling power plants with high sulfur coal in compliance with emission standards, Energy 9(3), 265 - 274
43. ZELL, R.A. (1983). Düstere Aussichten - Und ewig stöhnen die Wälder, Bild der Wissenschaft 12, 96 - 102

SOIL DEGRADATION IN A NORTH EUROPEAN REGION

Jens Tyge Møller
Institute of Geology
University of Aarhus
Denmark.

SUMMARY

Degradation problems in Northern Europe are rather different from the desertification in dry climate regions. However, the problems can be rather serious, particularly because the same soils have been used for farming for several centuries. So degradation is mostly connected with soil structure and farming methods. Owing to the normally favourable climatic conditions the most is made of the farming opportunities, but on sandy soils partly on the expense of soil stability. Illustrated by Denmark as an example this paper looks at materials and processes in addition to measures being taken against degradation.

1. INTRODUCTION

In humid regions, unlike dry climate regions, degradation is almost exclusively the result of soil structure, wear and tear and more inappropriate farming methods. Structurally, sandy soils are most liable to move with the wind. The finest particles can be removed completely and drifting sand cover vegetated areas. Owing to their cohesion clayey soils are more stable and, if clayey soils are eroded, the loss of fine particles will be negligible in comparison with the total amount of soil, but clayey soils, improperly cultivated, can be damaged. Farming methods, choice of crops and measures against wind erosion are of great importance and thus the occurrence of damage, caused by wind erosion, can depend on the farming tradition.

Within the Scandinavian countries the farming land in Finland, Norway and Sweden are to a high degree covered with woodland, preventing wind erosion. In Iceland volcanic deposits, not covered with vegetation, may suffer from wind erosion. Large farming regions are found in Skåne in Sweden and in Denmark. What follows is based on research and experience in Denmark, where wind erosion damage has occasionally developed to become a real desertification problem.

2. CLIMATE

If the soil surface was left to itself, very few effects of wind erosion would be seen. Any damaged surface would probably soon be covered again with vegetation. The Danish climate is rather humid (Table I). The annual precipitation varies between 500 and 800 mm with the lowest values in a zone from Store Bælt to Kattegat.

The wind climate is in this connection important. Strong gales mainly occur during the winter. The predominant wind directions are from the sector SW-NW, but during the spring easterly winds are frequent. During periods with easterly winds sea level in the Kattegat and the Baltic is rather low. Wide sandy flats are exposed and can act as sources for sand-drift when they dry out. Further, easterly winds frequently blow during cold

	Temperature			Wind			Precipitation mm		
				Veloc.	Direct.				
	1	2	3	4	5	6	7	8	9
	Mean	Max.	Min.	>4	E	W	Mean	Max.	Min.
January	0.0	3.9	-8.3	40	33	46	55	108	10
February	-0.4	4.1	-8.3	37	34	45	39	146	0
March	1.6	6.6	-4.1	32	38	38	33	82	6
April	6.1	9.1	1.3	26	30	42	39	123	2
May	11.0	14.2	6.5	22	36	36	38	104	2
June	14.5	17.7	11.8	21	23	52	48	144	1
July	16.6	19.9	14.2	20	21	56	74	205	11
August	16.3	20.5	13.0	21	25	50	80	207	2
September	13.2	16.6	9.9	28	26	52	72	189	2
October	8.8	11.7	4.6	36	34	43	70	180	2
November	5.0	8.7	1.6	38	37	43	60	139	6
December	2.4	5.4	-1.1	38	30	47	54	118	5

Table I. Danish climatological recordings from the period 1931-60. 1 and 7, average values. 2 and 8, greatest average recorded. 3 and 9, lowest values recorded. 4, wind velocity in % of all observations above wind force 4 (5.5-7.9 m/sec). 5, winds from the sector NE-E-Se in % of all observations. 6, winds from the sector SW-W-NW in % of all observations.

frosty periods leading to a weakening of the soil structure. During the spring large parts of the farming land are left bare owing to farming methods and choice of crops and thus storm damages very often occur during the spring, before the vegetation cover is complete.

Owing to the storms during spring, when considerable areas are left bare, the nature of surface is of great importance. Furrows left by ploughs and harrows, dry straw and remains of weeds are reflected in the absence of sand drift (Fig. 1). As soon as the fields are covered with vegetation in May-June, erosion stops, but fields damaged by sand movement have to be re-sown, in stormy periods several times.

Fig. 1. Field in Jylland seen from the north during a sand storm in March 1969. The sandy soil was frozen and on bare surfaces sand in saltation just moved across the field. In the grass in the background and among the straw in the foreground sand was deposited owing to the surface roughness.

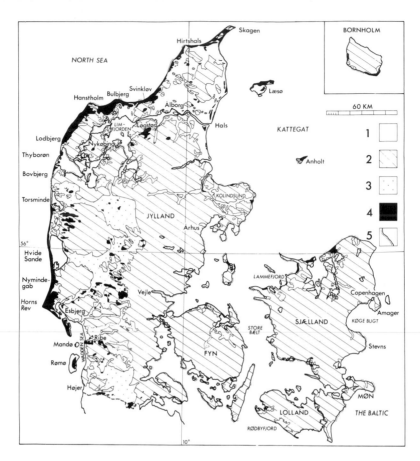

Fig. 2. Main features of the surface deposits in Denmark. 1, uplifted sea
floors including salt marsh at Højer and Ribe. 2, glacial deposits. 3, fluvial deposits. 4, blown sand and dunes. 5, coastal cliffs.

3. SURFACE DEPOSITS

Apart from the bedrock areas in Finland, Norway and Sweden the soil
parent materials in the Nordic countries have predominantly been created by
glaciation. Recent geomorphological processes in Denmark are due to rivers,
waves, currents and wind. Large parts of southern Sweden and almost the
entire surface of Denmark are covered with glacial and fluvio-glacial deposits (Fig. 2). In the North Sea region and the Scandinavian peninsula vertical movements of the land surface in relation to sea level have exposed
large areas of land. Owing to wave and current action much of the fine sediment has been washed out and mainly sand and coarser materials are left back.
Owing to the action of wind, waves and currents fluvial and glacio-flu-
vial deposits are well sorted in Denmark. The size classification appears
from Table II. In Denmark medium sand can be found on coasts facing the
Kattegat and fine sand on tidal flats. Fluvial sand is frequently coarse
sand between 1 and 2 mm.
In sheltered waters and in particular in sheltered parts of the sea
with tidal movements, fine sediments are found. As a whole most surface sedi-

Diameter in mm	Name
2 - 1	Gravel (very coarse sand)
1 - 0.5	Coarse sand
0.5 - 0.25	Medium sand
0.25 - 0.10	Fine sand
0.1 - 0.05	Very fine sand
0.05 - 0.005	Silt
Below 0.005	Clay

Table II. Grain size classification according to United States Bureau of Soils. In relation to Danish conditions the limit between fine sand and silt is a little too low. Normally the limit 0.063 mm is used in Denmark. Bagnold (1941) used 0.01 mm as limit between sand and dust.

ments in Denmark can be eroded and transported by wind, but deposits of well sorted sand are most liable to suffer from wind erosion. In Fig. 3 the clayey soils are shown as residuals, because wind erosion can occur on so many soil types that subtraction of soils only a little affected by the wind, seems to be the most relevant way of illustration.

Fig. 3. Clayey soils and other soils not liable to suffer from wind erosion are black. In addition to clayey soils peat bog, salt marsh and permanently wet areas normally resist wind erosion.

4. THE WIND

The degradation problems cannot be solved without knowledge of the wind's behaviour, in particular the velocity variation with height. Above a surface of some roughness there will be a smooth velocity transition between the undisturbed wind, far above the surface, and the wind close to the surface and thus affected by roughness.

The wind regime above a surface of some roughness has been closely investigated by R.A. Bagnold (1). Owing to turbulence wind velocity does not increase linearly with height above the surface and the wind regime cannot be described with a single wind measurement. The increase of wind velocity with height above the surface can be described logarithmically as a straight line (Fig. 4), and now the wind profile can be defined with two velocity measurements. The symbol for velocity is v or V. However, it is very important to stress that V_* in the following is the velocity gradient, not just a velocity.

If v_z is the wind velocity at the height z, $v_z = 5.76\ V_*\log \frac{z}{z_o}$ for a fixed sand surface. In relation to the sand surface, z_o is the height at which the wind velocity is zero and thus z_o is an expression for the surface roughness, approximately 1/30 of the surface irregularities such as ripples or grains. If the sand is moved by the wind, $v_z = 5.76\ V'_*\log \frac{z}{z'_o} + V_t$, where V_t is the threshold velocity as measured at the height z'.

$V_* = \sqrt{\frac{\tau}{\rho}}$ and $V'_* = \sqrt{\frac{\tau'}{\rho}}$, ρ is the density of air and τ is the drag. V_* and τ are measured above a fixed rough surface, V'_* and τ' above a surface with moving sand. In a logarithmic plot (Fig. 4) V_t, V'_* and V_* can easily be determined graphically. v_z has to be measured at two points to determinate the wind profile. Wind profiles above a rough surface clearly show a trend to intersect close to a point, O and O' respectively. O is situated where the air movement starts, O' where the grain movement starts. Thus the line connecting O and O' is the gradient, the wind profile, above which the grain movement starts, by Bagnold named the impact threshold, $V_* = V'_*$. Owing to the winds effect on a grain's surface the impact threshold varies almost as the square root of the grain diameter.

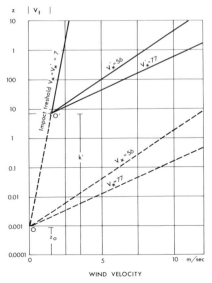

Fig. 4. Schematic outline of the wind velocity distribution with height above surface of some roughness (Bagnold 1973). The gradient is determined as the abscissa divided by the ordinate. If the ordinate is increased 10 times, V_* is the increase of the abscissa divided by 5.76. The hatched lines are gradients above a fixed sand surface, the thick lines above a surface with moving sand.

WIND VELOCITY

It appears from Fig. 4 that at the same height, the wind velocities
are greater above a fixed sand surface than above a surface with moving
sand. Further, it can be seen that the height z_o' corresponds to the threshold velocity V_t. Below this value no grain moves at all. z_o and z_o' are
values essentially to the stability of a soil surface with loose sediments.
The definition of z_o is rather difficult owing to the definition of the
surface. In Table III is shown some values for z_o for different surfaces.
It is essential to increase the effective surface roughness to ensure a
wind velocity gradient below that causing particle movements on the surface.
Surface stability depends on reducing wind velocity near the surface. Surface roughness can be created in many ways. Vegetation is a very important
factor, but patterns in the soil surface can also increase the roughness.
Humidity is no guarantee against sand movements. If the wind velocity is
great enough, even wet sand will start moving. On a beach sand drift can be
seen just above the upwash of waves. On the other hand, owing to cohesion,
humidity will of course reduce the movements of sand to some degree.

Ice and mud flats	0.01	mm
Short cut lawn (1 cm)	1	-
Flat snow field	2.3	-
Downland, thin grass up to 10 cm	7	
Thick grass up to 10 cm	23	-
Thick grass up to 50 cm	90	-
Wet sand on a beach	0.002	-
Beach sand with stones	0.1	-
Built-up areas	7-10	m

Table III. Values of the roughness parameter, z_o. The 5 values named
first, are from Liljequist (4), the rest from Kuhlman (7).

6. PARTICLE TRANSPORT

Particles are transported by the wind in three different ways. The
finest particles are picked up by the wind and moved in suspension (Fig. 5).
Bagnold suggests 0.01 mm as the limit between "dust" and sand and with that
the limit for suspended transport; but of course the limit depends on the
wind profile. Particles in suspension will only settle out with a change in
conditions such as a shift in temperature, in chemical composition (Fig. 6)
or rain. Sediments transported in suspension are normally lost from the
region where they were picked up. During strong gales stratified sediment
transport can be seen. The sand drift forms a white layer up to approximately one metre above the surface. The transport of very fine and light particles can be seen as a brownish cloud 20-100 m above the surface. The fine
particles are sorted out and deposited far from the place where they were
picked up and can easily pass over a sea such as the Kattegat. In calm
weather air movements, caused by heating of the soil surfaces, can pick up
fine particles and lift them to a height at which they can be transported
further, even by light winds.
The two other means of transport are mutually connected. Small sand
grains are lifted by the wind, because small, regular features in the apparently smooth sand surface leave the grains partly exposed to the wind.
Even a surface of well sorted grains will offer some irregularities, because
the sand grains form rings of 5 grains with a grain in the centre a little
above or below the ring.

Fig. 5. Sand and dust lifted by the wind during a strong gale.

Fig. 6. Road, only passable at low tide, to an island in the Danish Wadden
Sea. The marine sand, normally white, has been covered with organic matter
transported in suspension from further east during a sand storm in March
1969. The suspended material probably settled down when it came into contact
with salty air masses.

The smaller sand grains jump following trajectories depending on the wind velocity and the grain size. Bagnold introduced the name saltation for the jumping movement. A saltating grain may hit another grain of almost equal size and send this grain into saltation (1 and 2). A saltating grain can also hit the sand surface almost at right angles and remain motionless. A third possibility is an impact on a larger grain, too great for saltation but too small to remain unaffected by the impact. Such grains, of a size up to 7 times the diameter of the saltating grains, will move a short distance. This movement has been named surface creep.

The transport form saltation/surface creep is more complicated than just a movement of pushing and pushed grains. Saltation leads to a disturbance of the entire sand surface and the grains lifted above the surface are transported in the direction of the wind.

The formation of ripples will change transport (1). Normally the saltating grains will not hit the lee side of a ripple and consequently creeping grains will stop close to the crest of the ripple. Thus a characteristic feature of the saltating-creeping movements is the coarsest grains situated on the top of a ripple. Here they contribute to an increase in roughness and improved stability.

7. SORTING

Owing to the nature of particle movements a sorting of drifting sand will take place. The coarse grains are moved more slowly and can only move if they are occasionally hit by small grains. The small grains are faster moving owing to the saltation. The differences in transport mechanisms and in speed lead to formation of the ripple pattern. At very low wind velocities nothing will happen. At increased velocities ripples are formed, depending on the grain size, the sorting of the available sand and the wind profile. At further increased velocities the ripple pattern disappears and then, at still greater velocities, large ripples are formed.

The combination of the two types of transport is a sorting process, in time leading to a stable surface, because coarse grains are left on the ripple crest and the increase in roughness prevents further erosion. After a strong wind so much material can be removed that the surface, subject to wind erosion, is left covered with a natural pavement of grains not movable (Fig. 14). Such surfaces are resistant to wind erosion and fine grains are brought to the surface. Thus the sorting process leaves the eroded surface as a sedimentary residual, characterized by removal and a diminished sorting of the surface deposits.

Surface creep can cease gradually owing to changed grain size distribution with the largest saltating grains too small to move the smallest grains liable to creep. If the grain size distribution is very narrow, the grains will either saltate or remain motionless. If all grains are saltating, ripples will not appear because the ripple formation depends on the combination of transport forms. Wind erosion leads to the disappearance of material fine enough to be in suspension. In accumulation areas the deposits are sorted. Eroding areas are on the other hand left with a cover of poorly sorted coarse grains. Soils liable to suffer from wind erosion and transport always reflect the latest wind regime.

Mineralogically quartz sand predominates in Denmark. All types of sediments contain more or less sand and can be sources for aeolian transport: marine sand, morainic and fluvial deposits. Even boulder clay can nourish sanddrift during a strong gale, particularly if the soil surface has dried and cracked during a frosty period. Fig. 7 shows some critical wind profiles expressing the impact threshold (-gradient), V_{*_t}, at which grains of different sizes will start moving. A value of $V_{*_t} = 0.5$ m/sec is associated with a wind velocity of 14 m/sec measured at a height of 2 m above the soil surface. The importance of the combined transport process, saltation/surface creep, to the sorting process still has to be borne in mind. During the entire transport process the greater grains, able to saltate, must be together with grains, small enough to be moved by the impact of saltating grains.

The importance of grain size appears from Fig. 8 indicating typical impact thresholds for quartz grains. Grains with a diameter below 0.1 mm need larger values of V_{*_t} to start moving owing to the increasing importance of cohesion. Impact thresholds, V_{*_t} are difficult to measure owing to the varying conditions of roughness, shelter effect and wind velocity. Because it depends on local wind conditions, the impact threshold cannot be based on wind statistics from a meteorological station. Furthermore grain size and grain movement affect the value. Therefore the wind velocity has to be measured directly in the area in question. The behaviour of organic matter is connected with much uncertainty beacuse of the badly known shape factor. Sand grains of quartz, transported and rounded by erosion, are almost spherical, and all measurements and calculations are done under the assumption that the grains are spherical. However, mica grains are normally shaped like thin discs. With regard to grains of organic matter generalizing is impossible, because the buoyancy of organic matter particles is frequently great owing to their large surface areas and transport in suspension can involve rather large grains.

The interdependence between saltating and creeping grains under the same transport conditions appears from Fig. 9. It can be seen that a wide section of a grain distribution will move, first as surface creep and, at increased wind velocities, in saltation. The changing sorting conditions lead to a difference in the morphological effects. Fig. 10 shows wind velocity profiles required for suspended transport of quartz grains. A corresponding diagram for organic matter would be very complicated owing to the unknown shape factor.

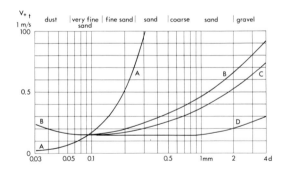

Fig. 7. Critical wind velocities expressed as the velocity gradient, the impact threshold, V_{*_t} in the height interval 0-10 cm above the soil surface (6). A, the lower limit of suspended transport. Above this graph a grain, already lifted from the surface, will remain suspended. The settling velocity in air for a grain with a diameter of 0.10 mm is 10 mm/sec. B, fluid threshold, the lowest velocity gradient at which sand movement will start owing to wind alone. Below this graph grains, not already lifted into the air, remain motionless. C, impact threshold for saltation grains. D, lower threshold for surface creep.

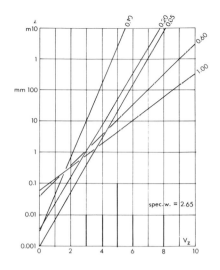

Fig. 8. Impact thresholds indicating the lower limits of velocity gradients needed to start the movements of quartz grains (6). Owing to their cohesion small grains below 0.10 mm need a rather great velocity gradient before they start moving.

The sorting process is of great importance to sandy soils, which can be seriously damaged because the suspended materials, essential to soil structure and to the occurrence of nutrients for the plants, will disappear. Blown sand cover and destroy the vegetation. Even richer soils, far from the source of the sand drift, may be covered. Left alone, a soil affected by wind action will become stabilized with time because the sorting processes produce stability. Unfortunately farming methods lead to disturbance of the surface and consequently the processes start again whenever a surface is left bare without sufficient roughness, either due to texture or to vegetation. Thus the control with sand transport and degradation is a matter of maintaining surface roughness.

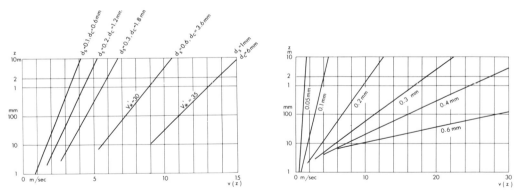

Fig. 9. Wind profiles indicating, at a given height, the minimum velocity required for maintainance of saltation and surface creep of quartz grains. d, diameter. c, surface creep. s, saltation (6).

Fig. 10. Wind velocity profiles indicating, at a given height above surface, the minimum velocities at which quartz grains can be transported in suspension (6).

8. QUANTITIES OF TRANSPORT

Quantitative measurements of transport are rather difficult, owing to the nature of the processes. Apparently the degrading of a surface is a rather slow process, but it can continue for centuries. Transport rates on river beds and on beaches can be determined by levelling, but such measurements are inapplicable to aeolian sand transport. An average ablation of just 10 mm, hardly measureable by levelling, during a storm will probably destroy the crops, and a field cannot stand an ablation of that size for many years. However, that ablation rate will result in a transport of 100 m^3/hectare.

Much research has been done on transports rates. Bagnold (1) has shown that the transport rate is directly proportional to the third power of the excess of wind velocity over the threshold velocity, V_t. The transport rate can be calculated from the function

$$q = \alpha C \sqrt{\frac{d}{D}} \cdot \frac{\rho}{g} \cdot (V - V_t)^3, \text{ where } \alpha = (\frac{0.174}{\log z/z_o'})^3, z = 1 \text{ m,}$$

$z_o' = 10$ mm, $V_t = 4$ m/sec, $C = 1.8$. The specific weight of air is ρ and ρ/g $= 1.25 \times 10^{-6}$. D is a standard grain size of 0.25 mm and d is the grain size of the sand in question, in the present case $d/D = 1$. At a wind velocity of 16 m/sec, the sand transport across a section of 100 km, a realistic lane width in Jylland, will exceed 2×10^6 m^3 in 24 hours.

Bagnold measured the transport of dry sand in a wind tunnel. Kuhlman (6) measured the transport over a firm, moist sand surface in a coastal region. The winds and the grain sizes were the same, but V_t was different and, owing to the third power, the effective wind velocity is of great importance. Still the values from the two investigations are of almost the same magnitude, if the difference in sand is taken into account. Probably determination of transported quantities of sand require a better knowledge of the velocity sector in which the transport starts. Bagnold (2) has shown a sudden start of transport when the threshold is reached, and from the threshold and upwards the log-log plot of transport against wind power is a straight line.

9. HISTORY

Through the ages wind erosion and sanddrift have occurred in Denmark. Layers of blown sand are found in many places, frequently overlaying old deposits of top soils with greater content of organic matter. Settlements, moved owing to sanddrift and later re-established on the old site, but separated from the first by layers of blown sand, can be dated to the Stone Age. In historical time, written sources tell of damage done by sanddrift. In particular along the west coast of Jylland and in northern Sjælland close to the sea, regions damaged by blown sand can be found.

Sanddrift and wind erosion were not restricted to regions along the shorelines. In several places inland dunes can be found in Jylland (Fig. 2). West of the town Vejle local people can point out the site of an old village and even find the floors of farmsteads, used by their ancestors. Today the entire region is a dune landscape, and the farmsteads are situated some kilometers outside the dunes. The causes for occurrence of sanddrift and wind erosion in historical times cannot be explained satisfactorily. However, damage caused by too much grazing by cattle, inappropriate cultivation of farming land and cutting of forests for timber and fuel must have been of great importance. So was the establishment of colonies of smallholdings in the 20th century, when marginal sandy soils were reclaimed.

Nowadays two obvious causes for sand drift and wind erosion can be found. Owing to the increasing size of farming machines, fields of large size are preferred. Further, the choice of crops results in bare surfaces during the spring. Finally Danish farmers are very fond of smooth, nice-looking fields with surfaces like flower-beds. It is difficult to convince a farmer that in fact he pays for destruction of the soil through his extra efforts. Frequently the last rolling after sowing is useless and directly harmful, because the roughness is so low.

Preventing damage from holiday traffic is almost hopeless, because most tourists have very little understanding of the problems. Except in Denmark, sandy beaches are rare in northern Europe and Danish beaches are very attractive to tourists. In Denmark many visitors meet the sea for the first time in their lives and have no knowledge how vulnerable vegetation on blown sand can be. Further, lots of holiday cottages are built, especially in sandy regions. They are close to the sea, the land was cheap and local communities want as many holiday cottages as possible to get hold of the taxes and enjoy the increase in local business. In some regions several kilometres of coastal belts are crowded with cottages and hotels built directly facing the sea. Consequently no space has been left for measures against sanddrift.

10. MEASURES AGAINST SANDDRIFT

In Denmark the importance of sanddrift is reflected in the measures taken during historical times. From 1539 an act provided that vegetation should be protected against any destruction, because at that time the vegetation cover, in particular on poor, sandy soils, was utilized for fodder, fuel and roof thatching. Plantings against sanddrift and also the establishment of wind fences started as experiments in 1824. According to statutory instruments, issued in 1792, the dune landscape near coasts has been managed by a special authority, now the directorate of forestry. In case of uncontrolled sanddrift in such regions the local forest supervisor, who has the right of conscripting local people for protection works, can take over management.

At the end of the 18th century only 4% of Denmark was forest-clad owing to the use of wood for fuel and ship construction. To prevent sanddrift in northern Sjælland, after some farming land had to be abandonned, plantations of conifers were established about 1770. From then the area of woodland increased, especially during the last 100 years, and now the percentage of woodland is approximately 11%. Large parts of sandy soils in Jylland are now forest-clad owing to the soil being unfit for agriculture. After the Danish-Prussian war in 1864 efforts were made to utilize as much as possible of the remaining area. Great parts of the sandy soils, not planted with conifers, were reclaimed and it was soon realized that sandy soils could not be cultivated without sheltering and, in addition to the plantations, wind fences were planted. Now this combination forms a surface roughness ensuring an acceptable soil stability.

At the end of the 18th century the Danish government called for German foresters to re-establish woodland and the new skilled foresters introduced many species of conifers. At that time almost all natural growing conifers has disappeared long ago, probably owing to climatic changes in the past. Most of the introduced species were very suitable for Denmark, but unfortunately some of the least suitable were selected for wind fences and plantations on sandy soils in western Jylland. Thus Mountain Pine (Pinus mugo) was planted as a pioneer tree along the coasts. It is able to grow in

Fig. 11. Sanddrift through a dying wind fence of White Spruce. The wind
came from the background. Near the ground openings between the trunks
caused an increase of wind velocity and owing to the limit of height the
velocity gradient was large. Sand, drifted through the fence, was deposited
on the road, supposed to be protected by the fence. During the strong gale
in March 1969 this road had to be kept open with machines, normally used
for snow clearance.

Denmark, but develops very poorly. White Spruce (Picea glauca) was used for
wind fences. Unfortunately this species is vulnerable to diseases. Further,
during growth trees lose their lower branches though they remain very
dense in the upper parts. Consequently, the dense growth leads to formation
of eddies on the lee side, causing increasing wind transport movements.
In the lower part of the wind fences the openings between the trunks cause
an increase in wind velocity as jet effects resulting in increased sand
drift (Fig. 11).

11. SHELTER EFFECTS

It has been mentioned that wind sorting of surface sediment leaves the
surface covered with a natural pavement preventing further sanddrift, except if the pattern is disturbed by cultivation or traffic. Consequently,
such stabilization only has a small effect in a farming region. However,
if the wind has a predominant direction, a surface structure perpendicular
to the main wind direction can be a sufficient protection against erosion.
The surface structure can be created in the shape of furrows left by agricultural implements or crops growing in rows such as potatoes.
As a remedy against sand drift, wind fences are more stable, but they
are very expensive, in particular if climatic conditions prevent the growth
of hedgerows, which is the case close to the shore. Apart from damage to
the dune landscape itself, the dunes can nourish sanddrift damaging other
regions. Owing to sand and salt from the sea, trees cannot grow close to
the shoreline. Tourists' wear and tear of the dunes is an important problem,
because the traffic hinders any measure against sanddrift. To reduce the

traffic, and at the same time increase the roughness, branches, normally of Mountain Pine, are stuck in the sand. Further, Marram Grass, (Ammophila are- naria) is planted in the dunes. Growth of this grass is stimulated by the sanddrift and the air's content of nutrients, but it cannot resist tramping and wave erosion.

Much research has been done on the effects of artificial and natural shelters. A wind fence consisting of a solid surface will create eddies on the lee side. Further, the shelter effect will only extend a distance down- wind approximately 20 times the height of the fence. Eddies will occur even if holes cover up to 35% of the surface. The maximum shelter effect is reached with 50% perforation, but the horizontal extent of the sheltered area is smaller than at the lower perforation rate (fig. 12). To avoid eddies and jet streams the perforation rate must be equally distributed on the fence. An effective reduction of wind velocity just above the surface can be measured over a distance of up to 30 times the height of the wind fence (3). Shelter effects, measured in a wind tunnel with wooden grills are shown in Fig. 13. The optimal rate of perforation is 35-40%. The great- est effects with regard to distance from the shelter are obtained with a high perforation rate, while a solid shelter without perforation has the greatest effect close to the shelter.

In addition to the effect of an isolated fence the total pattern of obstacles creating any roughness, will contribute to the entirety of shelter effect in the region. If fields, left bare for shorter periods, are spread in between the different shelters, it would be possible to control the wind erosion. On the other hand a minor change in agriculture methods, such as longer periods with bare soil surfaces, may destroy the effects obtained with shelters. Still worse is a major change in the degree of shelter, for instance if fences are removed because they occupy too much land. The con- sequence can be a total change in the shelter conditions in an entire

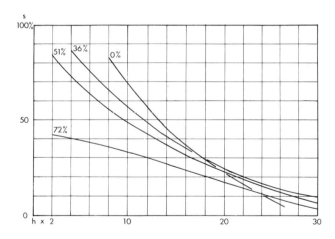

Fig. 12. Shelter effects of wooden grills (3). The percentages indicate the degree of perforation in the grill surface, s is the shelter effect and the abscissa is the horizontal effect of the shelter, expressed as a multiplicity of h, the height of the shelter. 72% is almost equal to an open hedgerow of White Spruce and B to a closed one. 36% is almost equal to the effect of a dense hedgerow of deciduous trees.

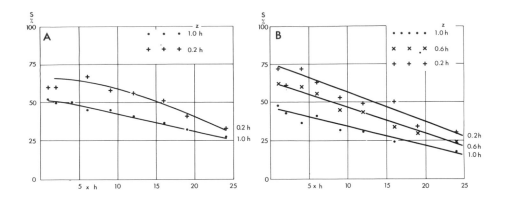

Fig. 13. Wind tunnel measurements of the shelter effect of a wooden screen with 48% perforation. The abscissa is the distance to the screen, measured with the screen height, h, as a unit. The wind velocity is measured in the height z. The roughness, z_0 is expressed as the height in which the wind velocity is zero. In A, z_0 = 2.2 x 10^{-4} x h, in B, z_0 = 18 x 10^{-3} x h. Owing to less turbulence in A the shelter effect is greater than in B (3).

farming region, just because a few farmers want large fields or want to cultivate land occupied by wind fences.

Fig. 13 illustrates the shelter effect of a wooden screen. The only difference between A and B is the values of roughness corresponding to a surface with crops (A) and without crops, B. The perforation rate, the average wind velocities and the wanted shelter effect have to be balanced, if a system of wind fences is to be planned. The most effective shelter is needed when the surface is bare, the roughness slight and the turbulence low (A). If the crop has grown so much that the turbulence is affected, the need for shelter is less.

Snow fences can catch sand as well as snow (Fig. 14) and from time to time cause troubles, because the drifting snow frequently moves together with drifting sand. Coarser grains can probably be transported in suspension carried by snow flakes. After deposition sand drifts are left when the snow has melted away. At very low temperatures the snow flakes grow rather hard and can to a certain degree grind sand grains free of the frozen soil surface. The combination of drifting sand and snow has caused problems to the railways when engines, moving into the snow drift to break their way through, hit sand drifts in stead.

Many efforts are being made to control sanddrift in Denmark and dune formation will soon be stopped. A mobile team of trained planters could be brought into action anywhere in Denmark for combatting sand drift by planting. However, formation of small dunes and other aeolian features cannot be prevented. Normally all traces of sand drift in the farming land are removed during cultivation. In old days farmers on the Wadden Sea island Rømø carefully removed the sand from the small fields and placed the sand around the field after each storm. Over time natural walls were formed enclosing the fields which are now situated several metres below the surrounding landscape.

Still many Danish fields are abandonned in result of sanddrift, though the greater part is planted with conifers and not just left unused. Traces of sanddrift are frequent in vegetated areas. Outside the period of growth

Fig. 14. Sand deposits on remaining snow behind a snow fence. The surface
in the foreground is covered with a natural pavement, pebbles left back
after the wind erosion removed saltating and creeping grains. Now the surface is stable at wind velocities up to those that caused the erosion. In
the far background a wind fence of White Spruce.

Fig. 15. Natural sand dikes created by deposition of sand in the vegetation
on a strip between two fields. Owing to ploughing, the limits of the dike
are rather well defined. Near the tree a primitive road had been kept open
by driving during the fields work. The road acts as a gate for the wind and
during a recent storm from west (the lower, left corner of the photography)
the sand drift found its way along the road and across the field in the
background. For a short period the outlines of the dike were blurred.

some straw and remains of weeds are left around wind fences, in the narrow
strips between fields and along road sides. The local increase in roughness
in a generally bare landscape leads to sedimentation during sanddrift in
the above mentioned places. Nobody is much concerned about the deposits
unless they prevent runoff from the roads. For ploughing fields are kept
as large as possible, and in this way natural dikes are built up along
wind fences, on strips separating fields and along the roads. Owing to the
slow growth and because the top is vegetated constantly, such dikes are
very stable and commonly exist with almost vertical sides (Fig. 15). Sanddrift dikes can be found in many places and some can be dated back to
prehistoric times.

Erosion during sanddrift removes fine particles, moves the sand and
damages plants. Deposition normally kills the vegetation. Sometimes entire
fields can be covered with uniform layers of sand. Denmark has a very long
tradition of the practical combat sanddrift and soil erosion. However, far
from all problems are solved and obviously there is still a lack of theoretical knowledge. Owing to long traditions and the humid climate wind erosion is in no way disasterous in the nordic countries. The problem is more
of a nuisance, at least in comparison with a dry climate region and only
small experiences and traditions for combatting desertification problems.

12. REFERENCES

1. BAGNOLD, R. A. (1973). The physics of blown sand and desert dunes.
    265 p. London.
2. BAGNOLD, R. A. (1979). Sediment transport by wind and water. Nordic
    Hydrology, 10, p. 309-322. Lyngby. Denmark.
3. JENSEN, M. (1954). Shelter effect. 264 p. Copenhagen.
4. LILJEQUIST, G. H. (1957). Wind structure in the low layer. In: Norwe-
    gian-British-Swedish Antarctic Expedition 1949-52. Scientific results,
    Vol. II, part 1 C. p. 188-234. Oslo.
5. KUHLMAN, H. (1958). Quantitative measurements of aeolian sand transport.
    Geografisk Tidsskrift 57, p. 51-74. Copenhagen.
6. KUHLMAN, H. (1960). Den potentielle jordfygning på danske marker.
    Geografisk Tidsskrift. 59, p. 241-261. Copenhagen.
7. KUHLMAN, H. (1976): Eolisk Geomorfologi. 36 pp. Geografisk Institut,
    Københavns Universitet.

INDEX OF AUTHORS